Digital Fabrication in Architecture, Engineering and Construction

Luca Caneparo

Digital Fabrication in Architecture, Engineering and Construction

Editorial
Antonietta Cerrato

Translation
Carolyn Winkless

 Springer

Luca Caneparo
DAD - Dipartimento di Architettura e Design
Politecnico di Torino
Torino, Italy

ISBN 978-94-007-7136-9   ISBN 978-94-007-7137-6 (eBook)
DOI 10.1007/978-94-007-7137-6
Springer Dordrecht Heidelberg New York London

Library of Congress Control Number: 2013946719

Printed on acid-free paper

Springer is part of Springer Science+Business Media (www.springer.com)

fabrication is certainly not a purely technological problem. It is indeed inseparable from cultural issues. Above all, it raises complex questions pertaining to the changing relations between the various professions and trades involved in architecture, engineering and construction. Ultimately, digital fabrication appears as a political subject, if we choose to define politics, after Bruno Latour, as the negotiated structure that regulates the relations between men and their actions.[9]

Harvard University   Antoine Picon
Graduate School of Design

---

[9] Latour B (2000) Politics of nature: how to bring the sciences into democracy. Paris. English translation (2004) Harvard University Press, Cambridge.

Acknowledgements

This book exists thanks to many contributions.

The subjects were discussed in depth with Liliana Bazzanella, Edoardo Piccoli and Benedetto Colajanni. Sadly, I did not manage to go through the whole book with Benedetto. I miss his acute and generous observations very much.

Whilst the author accepts full responsibility for any remaining errors, omissions or inaccuracies, Antonietta Cerrato and Valerio Pierantozzi patiently attempted to clarify ambiguities in the text, as did Francesco Guerra and Marco Broccardo with certain chapters.

I am indebted to all those who provided valuable technical documentation, listed in the index of Illustrations and Designs, particularly to Andrea Arnaldi of Associazione Italiana Macchine Tecnologiche e Utensili, Thomas Bock of the Technical University of Munich, Antonio Cossio of Woodengineering team, Alberto La Tegola of Stratex S.p.A., Matteo Simonetta of the Associazione costruttori italiani macchine e accessori per la lavorazione del legno, and Martin Tamke of the Centre for Information Technology and Architecture at the Royal Danish Academy of Fine Arts.

Special thanks go to Carolyn Winkless for her valuable translation work and her subsequent patient editing of my revisions.

I dedicate this English edition to Marilena, Paolo, and Silvia, who have been extremely generous in giving me the necessary time to complete the work.

Contents

Chapter 1
Introduction

Abstract A leading theme of the book is the evolution of the culture of architecture, engineering and construction (AEC), considered along the directions that innovation is taking with digital design and fabrication.

The Introduction addresses the different ways in which various countries have dealt with the issues of design, manufacturing, and construction, how the technical knowledge relates to design and to the ability to make, the different contributions provided by various techniques, and with them the meanings expressed by the architecture.

Digital technologies use the languages of information technology. Computer science deals with the study of these languages, and could be seen as either a branch of technology, providing the tools for progress, or as a branch of mathematics, therefore a methodology for the advance of scientific knowledge.

The book offers a perspective on innovation of digital technologies in design and construction that can help the reader to understand the context and direction of their evolution, and to interpret the requirements that drove their development, the challenges they were and are called upon to face that, in the end, contribute to shaping the technological culture in AEC.

The underlying theme of this book is the evolution of technologies for design and construction.

It grew out of my interest in studying the long term processes of technological innovation, the changes in the techniques used by design in dialogue with construction. A culture's ability to make and create is linked to its knowledge and understanding of making. In architecture this has a scale and breadth which, even merely in the number and variety of skills involved, offers a revealing cross-section of the scientific and technical disciplines of an era.

This study of how technical knowledge has been related to design and to the ability to make artefacts aims to shed light on the way in which different societies have addressed the problems of their time, to what extent they contributed to the development and what values they expressed with architecture. It is part of that research into technology and design, into the articulation of bodies of knowledge

L. Caneparo, *Digital Fabrication in Architecture, Engineering and Construction,*
DOI 10.1007/978-94-007-7137-6_1, © Springer Science+Business Media Dordrecht 2014

which, variously expressed by different sectors of a society, become "history" when they are taken up in widespread practice. Particularly the studies of Giedion, Mumford, Pérez-Gómez, Picon and Rossi have offered examples of methods which aimed to bring together a unified system of knowledge encompassing the techniques proper to each craft and discipline, to the historical position transmitted by technics with its methods and its ideology (Giedion 1948, 1954; Mumford 1934; Pérez-Goméz 1983; Picon 1992; Picon and Ponte 2003; Rossi 1970). Each study, from the common matrix of the analysis of the technics of an era, advances a way of thinking about their methodological organisation and their ends in a historical dimension.

From this common root, each investigation contributes to the aim of expressing an interdisciplinary reflection in combination with a personal interpretation.

The book ends with the innovation which inspires its title: digital technologies, a present-day stage in the evolution of technologies for design and construction. I hold that a historical perspective of a society's technical methods can help to give form, context and direction to innovations, especially to recent innovations in digital tools. Analysis of this contemporary innovation for construction and making artefacts offers a key to critical interpretation and the assessment of its capacity to resolve the problems which it is called upon to face.

> Construction is certainly an activity aimed at searching for ever more performances, from many points of view. But at the same time this systematic search for innovation is deeply rooted in the heritage of thousands of years of history which cannot be neglected. And it is precisely the capacity to keep this position of balance between novelty and trusted bodies of knowledge, between the future and the past, which distinguishes genuine technological innovation from technological innovation for the sake of it, often indifferent to its social and economic contexts (Campioli 2010).

I have borrowed a few paragraphs from the introduction of the book by Guido Nardi, *Tecnologia dell'architettura e industrializzazione nell'edilizia* [Technology of architecture and industrialisation in building]:

> The themes explored in this book derive from the assumption that, to fill the cultural vacuum in the field of technology and architectural techniques, it is necessary to connect the study of technical procedures to the development of philosophical and scientific thought and the political and economic use made of science and technics. It is equally necessary to analyse the specific problems of the sector by comparing them with the actual social, legal and technical situation (see chapters 3 and 4). In this way it is perhaps possible to reverse the current tendencies towards the increasing isolation and impoverishment of the technological and technical sector, which may be seen at several levels of analysis: conceptual, disciplinary and operational (Nardi 1980, p. 15).

The book crosses the boundaries of the different disciplines in the attempt to interpret the composite and varied contributions of the groups which have contributed and still contribute to the culture of building. It attempts to step over borders drawn at the birth of these disciplines with the development of the sciences at the beginning of the modern age.

These quotations have been selected because of references to ideas which, in retrospect, we know were important for several sectors of society. I have referred to them because they were able to express an emerging practice, making it become theory, to which researchers have attributed awareness of "history". The methodology behind this way of working is the conviction that technological progress, as a tool for responding to the needs of society, can be compared with (and partly driven by) the tools and ideologies, the methodological—and operative— knowledge, which came before and accompanied their birth and growth. This approach has the disadvantage of being unsystematic, which places the research on a different plane from the thorough procedures and systematic coherence of historiography. Nevertheless, it affirms a coherence within the practice of design which, measured against the reality of the problems to which it is expected to provide solutions, unifies and motivates the research. Furthermore, it constitutes an antidote to the illusion of inventing independent bodies of knowledge.

One of the aspirations is to offer a view, be it incomplete and biased, of the process of the specialisation of knowledge, which developed hand in hand with the development of sciences, technics and their languages. Since the dawn of the modern age, architects have often tried to bring together a unified body of knowledge and methodology, first around descriptive geometry and then around geometry as a specific discipline which "gives form" to mathematics, statics and mechanics:

> Geometry is the means, created by ourselves, whereby we perceive the external world and express the world within us.
> Geometry is the foundation.
> It is also the material basis on which we build those symbols which represent to us perfection and the divine.
> It brings with it the noble joys of mathematics.
> Machinery is the result of geometry. The age in which we live is therefore essentially a geometrical one; all its ideas are so orientated in the direction of geometry.
> Modern art and thought—after a century of analysis—are now seeking beyond what is merely accidental; geometry leads them to mathematical forms, a more and more generalized attitude (Le Corbusier 1987, p. xxi).

The first chapter begins with the recent history of mass production, looking for a parallel with the contemporary development of prefabrication in construction, comparing the developments in the United States with those in Germany. It considers the impetus given by the Second World War, which brought to fruition the process innovations developed by each country with the aim of fulfilling the demands of the war. These innovations thus played an important part in the outcomes of the conflict.

Contributions towards the end of the war were made by the prototypes of several innovative technologies, for example information technology with the first digital computers. The United States, with structural development policies and strategies considered in Chap. 2, placed themselves at the forefront of technology transfer Within 20 years they laid the foundations for the development of digital design and manufacturing in a broad range of industrial sectors, which would later be applied in the construction sector too.

In traditional crafts, a limited number of tools are used in a flexible and efficient way thanks to the acquisition of a deep level of manual dexterity and experience. In this phase, techniques favour versatility at the expense of efficiency, and as a result, the tools end up apparently simplified. But the case is rather that they are deliberately designed to function adequately in a variety of tasks, rather than optimally in a narrower range of tasks (Wescott 2001). One of the advantages of the technological strategy aimed at maximising tool versatility is that a smaller number of instruments reduces the time and investments necessary to make and maintain them.

The first industrial revolution introduced machines dedicated to specific functions, with the aim of increasing efficiency and reducing the know-how and manual skills required from the worker. The extreme development of this is the assembly line, in which the work is broken down into small stages which do not require specialised skills from the workforce, of whom each person repeats the same specific task. Digital manufacture develops flexible machine tools which integrate a wide and varied number of manufacturing processes. These include machining centres and robots, controlled by a new class of technical experts who possess the skills of the manufacturing processes and of the programming languages necessary to control them.

Subsequent chapters consider manufacturing processes according to the 'classical materials' used in construction.[1] Each chapter introduces the different manufacturing processes, considered via the study of buildings whose design and construction I consider significant for their digital languages and techniques.

Today several architects and theorists are eagerly looking forward to a recomposition around the *digital continuum*.[2]

To anticipate one of the conclusions, the case studies on digital fabrication and construction considered in this book demonstrate the possibility of a recomposition in the digital process which could reconnect different fields of knowledge and allow them to interact with one another through the language of the technical science, mathematics. Applied mathematics extends its application through the formalisation process of information technology, which develops the tools and the models beneficial to the purposes of science and technology. Information and communication technology puts into effect

> the standardization and automation of mathematical methods (and as such a reversal of the relationship of domination between pure mathematics and applied mathematics and, more generally, between theory and engineering). (Petitot 1985)

The redefinition of roles between theory and techniques applied to study and to the solution of problems began in the eighteenth century. Whilst the unprecedented possibilities of applied technics were being explored, intellectuals were trying to decide how technical knowledge should fit into the general structure of knowledge. The new arrangement was defined from the methodological foundations of the sciences, especially the mathematical sciences and, in the latter half of the century, from the developments in information sciences.[3]

> In 1967 Herbert Simon, Alan Perlis and Allan Newell at Carnegie Institute of Technology in Pittsburgh, argued that computer science was the study of computers [...]. As such

---

[1] From the title with the same name of the chapter in Torroja (1958, p. 24).
[2] See Chap. 3.
[3] Ibid.

computer science was a "science of the artificial" (Simon 1969). This new focus on computation further blurred the distinctions between science and technology since computation could be seen as either a human construction, and therefore technological, or as a branch of mathematics, and therefore a foundation of the sciences (Meijers 2009, p. 146)

The existence and the specialisation of different fields of knowledge precludes a synthesis. The language of the technical sciences allows the different disciplines to collaborate profitably to the pragmatic end of pursuing common objectives (Lyotard 1984; Harvey 1990), in our case, design—even if it does not achieve that unity between theory and practice which was lost when geometry became a branch within the mathematics, as a separate discipline.

References

Campioli A (2010) Innovazione e culture del progetto. In: Perriccioli M (ed) L'officina del pensiero tecnologico. Alinea, Firenze

Giedion S (1948) Mechanization takes command, a contribution to anonymous history. Oxford University Press, New York

Giedion S (1954) Space, time, and architecture; the growth of a new tradition. Harvard University Press, Cambridge, MA

Harvey D (1990) The condition of postmodernity. Blackwell, Cambridge. Italian edition: Harvey D (1993) La crisi della modernità. Il Saggiatore, Milano

Le Corbusier (1987) The city of to-morrow and its planning. Dover, New York

Lyotard JF (1984) The postmodern condition: a report on knowledge. University of Minnesota Press, Minneapolis. French edition: Lyotard JF (1979) La condition postmoderne: rapport sur le savoir. Les Editions de Minuit, Paris. Italian edition: Lyotard JF (1981) La condizione postmoderna. Rapporto sul sapere. Feltrinelli, Milano

Meijers A (ed) (2009) Philosophy of technology and engineering sciences. Elsevier, Amsterdam

Mumford L (1934) Technics and civilization. Harcourt Brace and Co., New York

Nardi G (1980) Tecnologia dell'architettura e industrializzazione nell'edilizia, 2nd edn. FrancoAngeli, Milano

Pérez-Gómez A (1983) Architecture and the crisis of modern science. MIT Press, Cambridge, MA

Petitot J (1985) Solo l'oggettività. Casabella 518:36

Picon A (1992) French architects and engineers in the age of enlightenment. Cambridge University Press, New York

Picon A, Ponte A (2003) Architecture and the sciences: exchanging metaphors. Princeton Architectural Press, New York

Rossi P (1970) Philosophy, technology and the arts in the early modern era. Harper & Row, New York

Simon HA (1969) The sciences of the artificial. MIT Press, Cambridge, MA

Torroja Miret E (1958) Philosophy of structures. University of California Press, Berkeley

Wescott D (2001) Primitive technology II. Gibbs Smith Publisher, Salt Lake City

Chapter 2
Origins of Industrial Production and Prefabrication

Abstract This chapter begins with the recent history of mass production, looking for a parallel with the contemporary development of prefabrication in construction and comparing developments in the United States with those in Germany. The United States pursued interchangeability of parts and standardisation in manufacturing from the time of its foundation onwards. This promoted and supported the initial stages of development and industrialisation, ultimately including mass production. Consideration is given to the driving role played by automotive manufacturing in developing the methodologies and the technologies of mass production and their role in the construction industry; and to the impetus given by the Second World War, which brought to fruition the process innovations developed by each country in response to the demands of the war and which played an important part in the outcomes of the conflict. While mass production tremendously increases the quantity of products and decreases their cost, the market was not particularly inclined to allow the mass production of housing. In the 1920s, when customers began to value other aspects apart from price, industries began to consider new manufacturing strategies— Ford himself for instance introduced the concept of flexible mass production. The construction industry stressed flexibility in design and manufacturing systems.

In the era of post-war reconstruction in Europe, the question of industrialisation in construction was addressed in response to the housing crisis, to update specific backwardness in Architectural, Engineering and Construction Industries (AEC), and to improve the quality more generally.

The Modern Movement made industrialisation a tool and a programme for pursuing a major reorganisation of the process, to offer the masses pouring into the cities a place to live which would conform to higher standards of housing and construction. More than half a century of experimentation may be briefly summarised by a number of projects and case-studies: Maison Citröhan and Maison Domino by Le Corbusier; Ready-Cut and Usonian Houses by Wright; General Panel by Gropius and Wachsmann; Dymaxion House and Wichita House by Fuller;

Fig. 2.1 McBean House built with the Marshall Erdman prefabrication system from a design by Frank Lloyd Wright

Magic House by Gunnison; House of Tomorrow by Keck; Aluminaire House by Kocher and Frey; Quonset Hut by Dejongh and Brandenberger; Case Study House by Eames; Deck House by Berkes; GEAI system by Marcel Lods; Maison Meudon by Prouvé; and Zip-Up Enclosures by Rogers (Fig. 2.1).

In Italy, new bodies and associations were founded to promote technological innovation in the construction industry; from 1957 the Centre for applied research into the problems of construction and the Italian association of prefabrication for industrialised construction, in 1964 the Italian association for the promotion of study and research in construction, in 1974 the College of construction industry technicians (Talanti 1975). Innovation was supported by standardisation activities and through the development of new standards in collaboration with interested parties, which was taken on by the Italian National Standards body (UNI) founded in 1928. Several Institutes were founded for certification, with the aim of organising and developing the quality of products and processes: in 1962, the Italian Institute for the Certification of Technical Competence in Construction; in 1988 the Institute for Certification and Quality Mark for Products and Services Used in Buildings; in 1993 the Institute of Quality Certification for Construction Firms and Services.

At the same time, exhibitions on prefabrication were organized, such as the Naples Fair of 1948 and the International Building Exhibition (Salone internazionale dell'industrializzazione edilizia) which, begun in 1965, is still the principal event related to the construction sector.

At the same time, and also as a result of these initiatives, further theoretical reflection and experimentation through case studies and projects were being pursued. In those years

> the debate on the industrialisation of construction concentrated on discussions between the proponents of the so-called closed prefabrication and the supporters of design for components. The first was identified by the use of elements which are linked together on the

basis of a predefined schema and in a single way. The second consisted of the acceptance of elements which could be connected together on an ad hoc basis, which therefore implies a certain freedom of assembly (Nardi 1986).

These two types of prefabrication have different cultural and technological origins.

2.1   Uniformity and Standardisation

The term building industrialisation refers to a whole range of different construction methods or procedures based, to a greater or lesser extent, on standardisation and the use of standardised materials (Nardi 1976).

The United States far-sightedly pursued standardisation in manufacturing from the time of its foundation onwards. The future president Thomas Jefferson, when he was still a minister in 1785, wrote from France:

An improvement is made here in the construction of the musket which it may be interesting to Congress to know, should they at any time propose to procure any. It consists in the making every part of them so exactly alike that what belongs to any one may be used for every other musket in the magazine. […] As yet the inventor [Honoré Blanc, A/N] has only completed the lock of the musket on this plan. He will proceed immediately to have the barrel, stock and their parts executed in the same way. […] He presented me with the parts of 50 locks taken to pieces and arranged in compartments. I put several together myself taking pieces at hazard as they came to hand, and they fitted in the most perfect manner. The advantages or this, when arms need repair, are evident (Jefferson and Holmes 2002, p. 62).

What Jefferson in his letter defined as "the making every part of them so exactly alike", led to the achievement of

standardisation and, closely connected therewith, the interchangeability of parts (Giedion 1948a, p. 49).

At the time, it was a complete innovation.

It is interesting to show how, in these early stages, interchangeability of parts did not in itself require mechanised manufacturing for rationalised assembly and production. However, Jefferson himself knew that mechanised manufacturing allows greater precision and repeatability than handcrafted production. Conversely, mechanised production does not imply interchangeability of parts, as shown by the firearms industry of the time in Austria, France or Great Britain.

In the decades following United States Independence, the American government systematically pursued interchangeability and standardisation, favouring and directly financing large commissions under the War Department for complete armaments or for individual parts, which had to satisfy stringent needs and criteria for interchangeability and precision. This decision established the requirements and furnished the resources for the rapid developments which followed, advanced by leading inventors, industrialists and managers such as Eli Whitney and Colonel

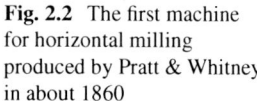

Fig. 2.2 The first machine for horizontal milling produced by Pratt & Whitney in about 1860

Samuel Colt. Jefferson himself, after attending a demonstration by Whitney on the assembly of interchangeable musket parts in 1801, wrote:

> Mr. Whitney is at the head of a considerable gun manufactory in Connecticut, and furnishes the United States with muskets, undoubtedly the best they receive. He has invented moulds and machines for making all the pieces of his locks so exactly equal that take 100 locks to pieces and mingle their parts and the hundred locks may be put together as well by taking the first pieces which come to hand (McLaughlin 1956, p. 133).
>
> Eli Whitney, inventor of the cotton gin, is held the first to have introduced interchangeability of parts in gun manufacture, in his Whitneyville factory (Giedion 1948a, p. 49).

Whitney is also regarded as the inventor of one of the first lathes in which the head carries rotating tools, processing a workpiece clamped to the bench. This machine proved to be particularly versatile in making parts both for manufacturing in general and specifically to produce other machine tools developed later, for example the cotton gin (Woodbury 1972). The development of an effective machine tool industry is considered a strategic factor in the industrialisation process (Fig. 2.2):

> This role was a dual one: (1) new skills and techniques were developed or perfected there in response to the demands of specific customers, and (2) once they were acquired, the machine tool industry was the main transmission center for the transfer of new skills and techniques to the entire machine-using sector of the economy (Rosenberg 1970a).

The inventor of standardisation and interchangeable parts, Honoré Blanc, was able to test the manufacture of standard flintlock mechanisms for muskets by the founding of a Military Arsenal in Vincennes. But when state support was reduced, his attempts to apply the methodology in order to supply the army through his private factory failed, together with his business. The role played by the USA Government, in particular the War Department, turned out to be fundamental in promoting and supporting standardisation and mass production in its initial stages of development and industrialisation (Mumford 1934; Hounshell 1987).

2.2   Mass Production

Nardi in *Progettazione architettonica per sistemi e componenti* [Architectural design for systems and components] considers:

> An industrialised project entails two fundamental presuppositions which contribute to defining the characteristics of the project itself: quantity, and a sequence of processes to make the product (Nardi 1976).

Among the pioneers of industrial prefabrication in the United States, Foster Gunnison made important innovations in technology and production. Gunnison applied the research of the U.S. Forest Products Laboratory for a stressed skin panel in plywood.

> The stressed skin principle was not new, except to housebuilding; the idea was simply to build the wall panel as a box girder and thus use the surfaces of the panel in such a way that they, as well as the framing members, would carry a major part of the load (Kelly 1951, p. 33).

In this way it became possible to reduce considerably the use of wood, lighten the structure and mechanise production. The Laboratory was to collaborate with Gunnison and other prefabricators in resolving the problems which mass production would encounter: deformation of the two sheet panels due to variations in moisture percentage, inter-panel condensation and dust deposition in patterns.

In 1935 Gunnison founded a company for industrial prefabrication, Gunnison Magic Homes, in New Albany, Indiana. The choice of site was due to the acquisition of the company Plywoods Inc., which soon allowed the production of structural panels in 4-ft by 8-ft units, 1 1/2 in thick (122 by 244 cm, 38 mm thick). At full capacity, the factory produced in 1937 twenty-four different models of houses from a modular standard wall component.

> Of the countries participating in the Economic Commission for Europe of the United Nations, the United States of America is the one with the longest history of achievements in the sector of dimensional coordination in AEC. As early as 1936 an American civil engineer, A.S. Bemis, published a book (The Evolving House) which proposed the generalised and systematic use of a basic module of 4 inches (10.16 cm). With the aim of further developing and promoting these ideas an association for modular studies was established, the efforts of which, among other things would open the way for the institution of a specific committee for modular coordination within the American Standards Association (Crespi 1967, p. 57).

Standardised modules had the aim of defining standard dimensional units which, as in manufacturing industry, would allow modular coordination between different producers and at the same time reduce the need to adapt parts at the assembly stage.

The modular wall was manufactured with pressure-fusion in order to attach the stressed-skin plywood panel to a metal structure of aeronautical design. The thickness of the panel was only 2 in (50 mm), including the insulation layer. Windows and doors were pre-mounted in the panel.

The whole process of production and assembly took place with an assembly line similar to the one tested and inaugurated in Henry Ford's Detroit Highland Park factory for making the revolutionary Model T car.

Gunnison adopted the theory of serial production, with the intention of capitalising on the popular image of the link between assembly line and mass production, and with the aim of demonstrating that houses can be produced in a factory with the production line methodology (Fig. 2.3).

William Blitzer (1938), in an article in *Architectural Forum*, analysed the relevance of Gunnison's contributions:

> This conveyor, probably the first used in the manufacture of houses, was a symbol of Gunnison's achievement, and pictures of it were reproduced very widely. It was taken as a sign that prefabrication had become a mass production affair, that it was following in the footsteps of the illustrious automobile industry and that, after many words had been exchanged on the subject, industrialisation methods were at last being applied to housing production. […] [Gunnison] had perfected prefabrication on a true mass-production, assembly line basis. […] Gunnison was the first prefabricator to use a moving production line.

Siegfried Giedion (1948b, p. 72) recognised how the production line is fundamental to the scientific management of industry.

> The assembly line is one of mechanisation's most effective tools. It aims at an uninterrupted production process. This is achieved by organising and integrating the various operations. Its ultimate goal is to mold the manufactory into a single tool wherein all the phases of production, all the machines, become one great unit. This division of production into small fractions of operations and their seamless integration is the key to contemporary mass production. The time factor plays an important part; for the machines must be regulated to one another.

David A. Hounshell (1987, p. 228), in a chapter entitled "The Ford Motor Company & the Rise of Mass Production in America", describes the visit of science journalist Fred Colvin to the factory in 1913:

> he was impressed by the way Ford engineers had concentrated on the principles of power, accuracy, economy, system, continuity, and speed, Henry Ford's elements of mass production. Noting that Ford manufactured over half the entire United States output of cars […] A complete Model T emerged from the factory every forty seconds of the working day.

The Highland Park factory visited by Fred Colvin shows Henry Ford's remarkable ability to identify the principles of production and the people able to translate them into a complete industrial system, all of which was the result of a project which had taken several decades. Highland Park was constructed expressly for the production of the Model T, utilizing processes and technologies developed in earlier Ford car models in conjunction with technologies that were innovative at the time.

Fig. 2.3 Operation of the Gunnison Magic Homes factory. (**a**) Sawing the plywood. (**b**) Sawing the uprights of the structure. (**c**) Assembling the panels. (**d**) Trimming the panels. (**e**) Multi-function press for gluing panels together. (**f**) Panel painting line

The principles of precision and organisation formed the basis of Ford's commitment to interchangeable parts, which, as we have considered, was one of the basic principles of industrialisation in the United States. Ford understood and emphasised on several occasions

> that to achieve mass production, interchangeability must be accurate and unique to allow the rapid assembly of the parts. A lot of manual work or adjustment must not be necessary, if you want to achieve great things (Hounshell 1987, p. 228).

As for the principle of continuity, in the production of the earlier vehicle, the Model N, machine tools had already begun to be placed according to the sequential operations on the various parts, rather than by the types of machine (for example one department for milling, another for pressing and so on). This is an important step towards the assembly line principle, called *moving the work to the men*.

In April 1913, the first assembly line was trialled in the manufacture of a magneto flywheel at Ford Motor Company; 29 workers each carried out a single, specific assembly operation, and passed the piece to the next worker in the line. The time taken to produce a magneto flywheel was reduced to 13 min, compared with the 20 min previously taken for the assembly of a complete alternator by a single worker. As the working of the line was perfected, and with the mechanisation of a conveyor belt for the workpieces, production was further improved until it reached the point where it took only 5 min to make a magneto flywheel (Hooker 1997).

Giedion, Hounshell and Mumford examined the contributions of scientific theories to mass production from the studies and experiments of Frederick Winslow Taylor and of Frank G.B. Gilbreth. In industrial manufacturing the assembly line allowed the continuity and the speed of the process to be monitored according to Henry Ford's principles. With the assembly line, in fact, the Ford engineers put into practice the principles of economy, system, continuity and speed, allowing them to integrate manual and mechanised processes into one system, synchronised with defined timings and assessed according to the production potential of the workers and of the machines.

On the basis of very positive experiences in the production of magneto flywheels, the Ford engineers began to plan the assembly lines of the Highland Park factory for the entire production of the new Model T. Henry Ford wanted to dedicate this factory exclusively to the production of the new Model. This gave its designers the ideal opportunity to install single-purpose machine tools, dedicated to a single manufacturing process, integrated along the production line (Raff 1991). The lines were planned according to the principles of economy and continuity, optimising the distances between the machine tools so as to reduce the time taken to move the workpieces.

The Ford engineers defined the space and distribution requirements for the building which was to host the new production lines: a block four storeys high. Production began on the top floor, to which raw materials were sent on rails. Manufacture proceeded downwards, moving along rails and conveyor belts to the ground floor, where the Model T came out complete.

The person assigned the job of designing the Highland Park buildings was Albert Kahn, selected because of his experience with construction systems using reinforced concrete, for which he invented a bar truss system. The construction system chosen (standardised frames in precast concrete) was innovative at the time, and worked to Ford's production principles: maximum clearance, to allow productivity, economy and speed of execution.

Unlike other industrial groups of the time, Henry Ford followed a policy of popularising his achievements. He advertised advancements through various publications and organised visits to the plants. In a short time, the principles and technologies developed by Ford spread to the whole of the United States and abroad. In Turin, they were taken up in Fiat's Lingotto factory, also built of reinforced concrete. Construction was begun by the Porcheddu firm, which had the Italian concession for the Hennebique patent, and completed by the group of firms coordinated by Giacomo Matté Trucco, which replaced Porcheddu during the course of the work (Olmo 1994; Olmo et al. 2003). The Lingotto factory put Ford's principles into practice through a modular construction on a 19 ft by 19 ft (6 by 6 m) grid. At the same time it made one exception to this: when it was already under construction, the project was modified to insert a new storey and the rooftop vehicle test track.

2.3 Rise and Decline of Mass Production at Ford Motor Company

Ford's principles, taken to the limit in the production of the Model T, enabled the production and sale of over 15 million vehicles in the USA. Certainly an epoch-making success, which has left its mark on methods of production, the market, mobility and even the popular imagination.

But after 13 years of uninterrupted production, the market for the Model T was saturated. In 1920 Ford Motor Company's market share had gone from over half to less than a sixth. In 1924, the retail price of the vehicle was approximately the annual salary of an average American, that is only 260 dollars compared with 569 in 1912.

The reasons for its success were also those for its decline: with the increase in standards of affluence, the Model T began to be considered old-fashioned. Elmo Calkins wrote that this car

did violence to the three senses, sight, hearing and smell (Gartman 2009).

In the 1920s, the average buyer began to value other aspects apart from the price. A consistent number of higher-range cars began to circulate on American roads, from Chevrolets to the luxurious Packards, Cadillacs and Lincolns. In popular imagination, the Model T became indelibly associated with the negative aspects of mass production: obsolete models, scarce attention to design and details, no adaptability or possibility of personalising the product.

The market and the consumers were demanding new products. Harvey N. Davis in the chapter "Spirit and Culture under the Machine" says that

mass production tremendously increases the quantity of useful things in the world, and decreases the cost of them (Hounshell 1987, p. 308).

He goes on to observe how mass production was a reflection, rather than a cause of early twentieth-century American society:

the technology a society wants is the technology it gets (Hounshell 1987, p. 308).

As early as 1925, Henry Ford began to consider the changes in strategy necessary for introducing new models. The design and production of the new Model A conflicted strongly with the very principles which Ford had put into practice at the Highland Park factory in single-purpose machine tools and in the optimisation of spaces and layout of the assembly line. The design and production organisation had been conceived and perfected for the mass production of one model. But it was soon evident that it would be impossible to adapt the existing plant to the production of a different model: a new factory was needed, suited to the new principles, with spaces not specialised for particular purposes. Rather, flexible spaces were needed, infinitely adaptable and expandable for any requirements.

Thus, hardly 3 years after the completion of a new six-storey building at Highland Park, Kahn began to plan the construction of a new factory at River Rouge, a suburb of Detroit. The new complex had to fit the new requirements of flexibility which Ford wanted: Khan designed a factory on a single floor, the so-called "horizontal factory", which could be expanded with further modules in any direction. The structure was of steel, covered by saw-tooth roofs to light the wide open spaces for manufacture.

The Fiat company, especially after a visit in 1926 to the Ford plants, began to consider replacing the Lingotto building with a new plant in keeping with the most recent innovations. The need for a new factory therefore gave birth to Mirafori, which would be located in the formerly agricultural area around Turin.

For Ford Motor Company, the demands of updating machine tools were to have enormous effects: in the production of the Model T, 43,000 machine tools were used, most of them specialising in a single function. For the production of the Model A it was necessary to update or to substitute more than half of these, at the price of enormous investments. Of the new machine tools, some remained single-use (for example lathes, milling machines, boring machines and grinding machines), because of their better performance in terms of time or of accuracy in processing. But a significant number were replaced by general-purpose machine tools, that is, versatile machines able to carry out several types of processes, including innovative turning centres.

As a consequence of these numerous and significant changes, Ford introduced the concept of flexible mass production.

2.4   Types of Mass Produced Houses at Gunnison Magic Homes

As for the mass production of housing, in 1937 Gunnison did not find a market particularly inclined to buy his products.

> Anglo-American experience with the introduction and diffusion of technology in the nineteenth century points strongly to the importance of the composition of consumer demand and the malleability of public tastes. The willingness of the public to accept a homogeneous final product was a decisive factor in the transition from a highly labor-intensive handicraft technology to one involving a sequence of highly specialised machines (Rosenberg 1970b).

From 1930 Lewis Mumford (1934) analysed the factors which had probably led to the lack of success in the mass production of housing. First of all, there is the minimal reduction of total cost: prefabrication in action concerns mainly the shell and the structure which does not constitute the main cost of a home. We should not underestimate the fact that in the USA the construction of single-family homes already commonly used widespread systems of prefabrication and semi-finished products, such as the balloon frame and platform frame systems.

A skilled carpenter and just one helper could erect a small one-story house measuring 14 by 20 feet [4,3 by 6,1 m] in about one week (Peterson 2008, p. 9).

According to Mumford (1934, pp. 314–315),

land, manufactured utilities, site-improvements, and finance call for a greater share of the cost than the 'building' and labor.

Durable or semi-durable goods such as homes are not compatible with the obsolescence and the replacement cycle which apply to fast-moving consumer goods. Mumford formulated a theory which he called the "Model T dilemma", that is the tendency of the mass production industry to attain premature standardisation, because of its own specific need for specialised machinery and the aim of the linkage of chain processes. The consequence for the industry is

to prolong the life of designs which should be refurbished (Mumford 1945, pp. 82–83).

Mass production needs to tackle the dilemma of whether to persevere with obsolete models or to allow the proliferation of changes of style, often an end in itself (Figs. 2.4 and 2.5).

In 1951 Martin Heidegger, in *Building Dwelling Thinking*, stated that the real housing crisis pre-dated the shortage of houses blamed on external causes such as war or demographic boom. Such a crisis, rather, was a consequence of the models adopted for construction. Heidegger writes:

That language in a way retracts the proper meaning of the word *bauen*, which is dwelling, is evidence of the original one of these meanings; for with the essential words of language, what they genuinely say easily falls into oblivion in favor of foreground meanings. [...]
  But if we listen to what language says in the word *bauen* we hear three things:

1. Building is really dwelling.
2. Dwelling is the manner in which mortals are on the earth.
3. Building as dwelling unfolds into the building that cultivates growing things and the building that erects buildings (Heidegger 1977, p. 326).

The interpretation of housing models understood by Heidegger agrees with that of his contemporary Mumford in considering mass production a deprivation, when it is applied to primary needs such as "dwelling", which involves more than simply fulfilling a use-value, as a shelter.

The real plight of dwelling does not lie merely in a lack of houses. The real plight of dwelling is indeed older than the world wars with their destruction, older also than the increase in the world's population and the condition of the industrial workers. The real dwelling plight lies in this, that mortals ever search anew for the essence of dwelling, that they must ever learn to dwell.

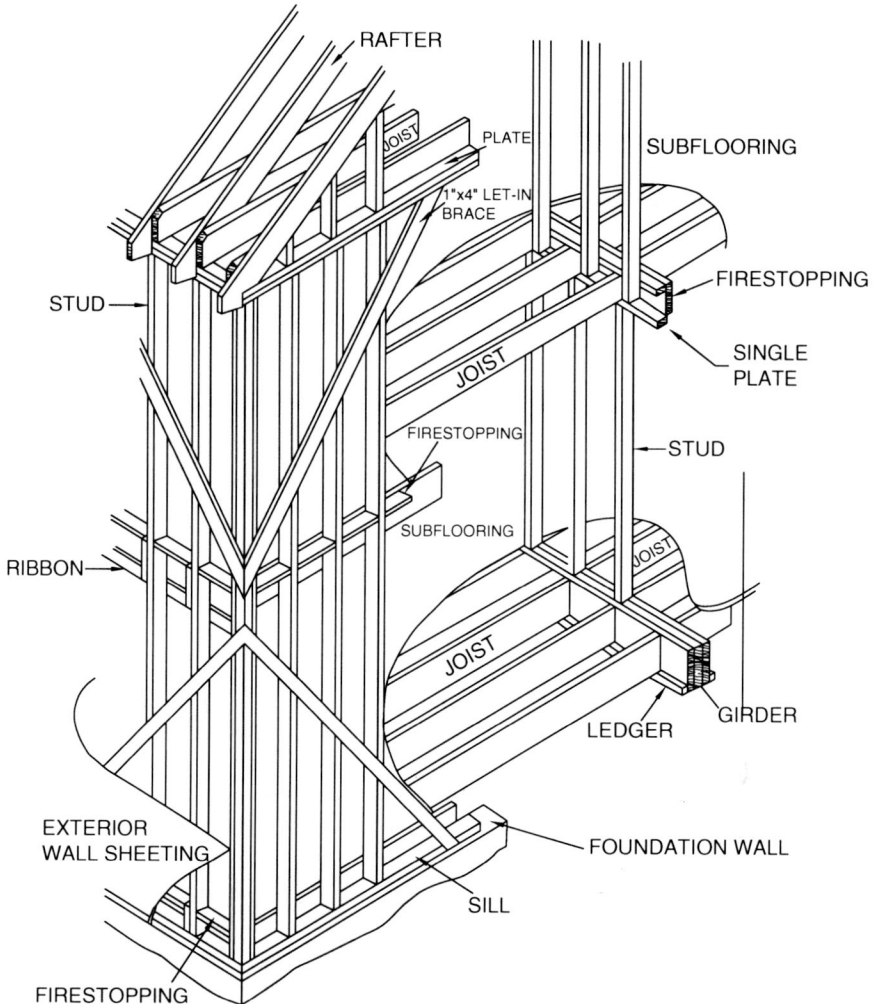

RAFTER

PLATE

SUBFLOORING

JOIST

1"x4" LET-IN BRACE

FIRESTOPPING

STUD

JOIST

SINGLE PLATE

FIRESTOPPING

STUD

SUBFLOORING

JOIST

RIBBON

JOIST

LEDGER

GIRDER

EXTERIOR WALL SHEETING

FOUNDATION WALL

SILL

FIRESTOPPING

Fig. 2.4 The balloon frame construction system

The word *bauen* (building or dwelling) implies a person's very being, not just residing; it

> also means at the same time to cherish and protect, to preserve and care for, specifically to till the soil, to cultivate the vine (Heidegger 1977, p. 339).

The reluctance of the market to accept the cultural models of mass production, particularly when aimed at durable goods like the home, was later aggravated by the Great Depression. Whilst the market had substantially decreed the end of

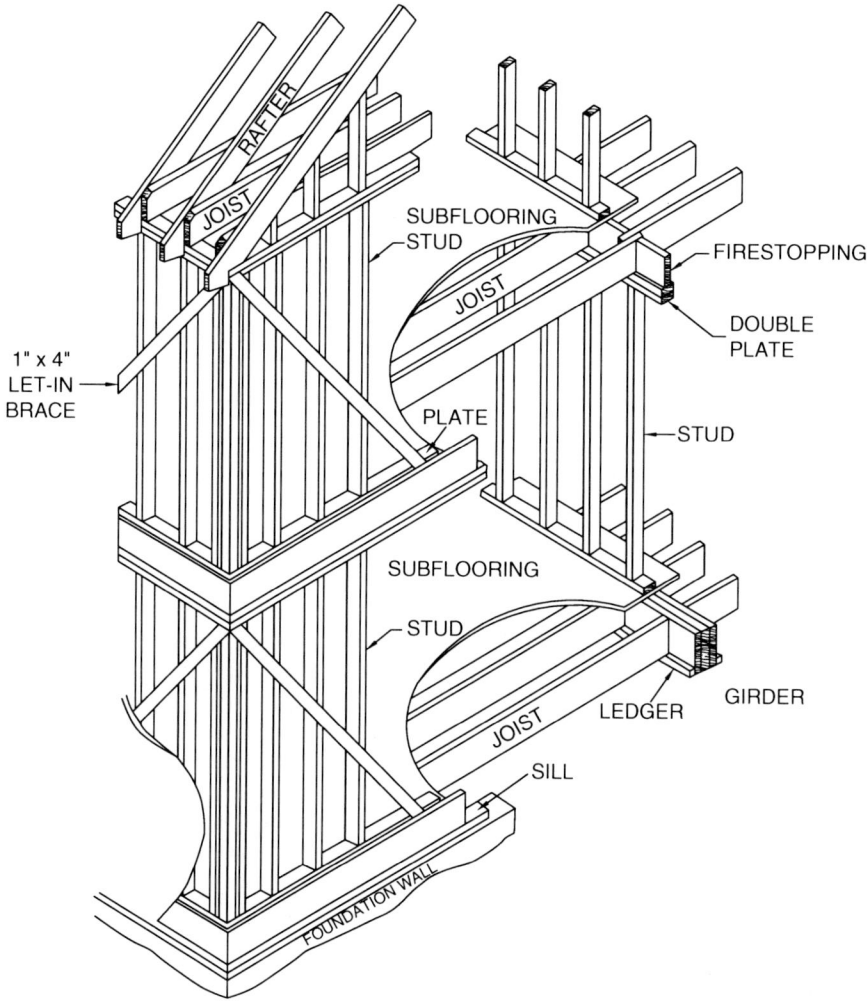

Fig. 2.5 The platform frame construction system

mass production, the beginning of the Second World War imposed an enormous manufacturing effort on a huge scale:

> Providing adequate housing for workers in the war industry is an integral part of our national assembly line. Houses must be rolled out progressively in parallel with the opening of factories. They must be planned to function as cogs in the gigantic flywheel which leads to victory. […] We cannot allow production to be hindered by the lack of adequate housing, nor can we support a programme of lodgings left to luck, which would only lead to the creation of barrack towns in the suburbs and the countryside immediately after the war. Good sense and energy associated with an understanding of the work before us must provide the necessary infrastructures for housing, to speed production for the war effort (Rosenman 1942).

At the end of the war, Gunnison's accounts summarised over 45,000 houses sold, most of these provided to the Army. To house workers in the military industry, the Lanham Act of 1942 ordered the provision of 70,000 prefabricated dwelling units. At the conclusion of hostilities, around 170 firms had produced a total of more than 200,000 prefabricated houses (Kelly 1951).

Only those industries organised for mass production would be able to satisfy the demands of standardisation and repeatability imposed by the conflict. Industrial production in the United States between 1940 and 1944 was increased by 15 %, with an initially slow growth due to the conversion of many industrial systems to arms production—this despite the heavy reduction in labour force, due to conscription, and its replacement with unskilled labour, often consisting of women in their first experience of manufacturing.

In 1934 Lewis Mumford believed war to be the real origin of the industrial revolution, rather than the invention of the steam engine.

> The spread of conscription and volunteer militia forces throughout the Western World after the French Revolution made army and factory, so far as their social effects went, almost interchangeable terms. And the complacent characterizations of the First World War, namely that it was a large-scale industrial operation, has also a meaning in reverse: modern industrialism may equally well be termed a large scale military operation (Mumford 1934, p. 84).

These considerations are particularly true of the Second World War, even more than the First. Perhaps it is not a coincidence that modular, standardised and mechanised production was born at the end of the eighteenth century in France in an arsenal, was rapidly adopted in the USA for military needs and on military specifications and found its greatest expression in the mass production of the Second World War. For the Allies, victory certainly depended, as people at the time themselves believed, on the capacity of the United States' mass production system to manufacture ships, aeroplanes, tanks and other systems of armaments in unprecedented quantities (Smith 1985; Milward 1977).

2.5 Automobile Industry in Germany

Germany also had ambitions in mass production and motorisation. From 1920, the automobile industry began to develop designs for the *Volksautomobil*, *Volksauto* and *Volkswagen*, intending to make cars accessible to the middle class. Candidates for production were either German industries, or those of the United States: Ford with its own subsidiary; General Motors from 1929 when it acquired the Opel Company.

The German automobile industries were interested in producing cars in a reduced number of models, not on a wide scale, since that would have required their updating to methodologies and technologies for mass production, with the industrial and investment problems well noted after the experience of the Ford Motor Company was divulged.

Fig. 2.6 Prototype Type 32 of Ferdinand Porsche's design for the Volksauto

In 1934 Hitler made a speech on the "People's Car", literally Volkswagen.

The will of the Government, in promoting the car, is not just to give an impulse to the economy and provide bread and work to hundreds of thousands of men, but also to allow ever wider masses of our people to have the possibility of using more modern means of transport.

Ford began to advertise its own Colonial Model as the people's car at a price of 2,000 RM. In 1935 it had to withdraw this advertisement on protests from the German association of car manufacturers (RDA). In 1936 Opel offered its own basic model, P4, at 1,450 RM.

Hitler decided to adopt the design of the Volksauto drawn up by Ferdinand Porsche (the famous designer of racing cars), with the aim of selling more than a million cars per year at a price of 1,000 RM, approximately the annual salary of a German worker. Previously, Porsche had proposed its design first to AustroDaimler, then to Daimler-Benz, who had declined because of the expected cost of production, which would have required a sale price well above what the average German citizen could afford (Fig. 2.6).

At the Berlin Motor Show of 1934, Hitler made a speech on the need for the standardisation of the automobile industry:

all the chief parts of any manufacturer's car should be interchangeable with those of all other manufacturers (Hill and Wilkins 1964, p. 272).

The standardisation which Hitler wanted imposed the rule that any part or component of a vehicle, irrespective of the manufacturer, should be interchangeable with those of others, be they Adler, Auto-Union, Daimler-Benz, BMW, Ford or Opel. The work of standardisation was defined by the Deutsches Institut für Normung (DIN), becoming a requirement for access to contracts and tenders.

The aim was threefold: firstly, to bring the industry as a whole under the influence of the Third Reich, forcing foreign products to conform to German rules, with a perspective of protectionism and autarky. This was particularly restrictive for Ford, which had always pursued a policy of interchangeability between its own parts even when they were produced in different factories. Secondly, it had the aim of preparing the German manufacturing system for the manufacture of armaments; the experience of the United States clearly indicated the benefits of an advanced level of interchangeability. Thirdly, it had propaganda purposes, demonstrating that the regime's control extended even to the industrial system.

By 1936, Porsche's design for the Volkswagen had required total investments of 1.7 million RM, yielding three alternative prototypes which would be subject to a series of extremely long and accurate tests. The purpose of these prototypes was to show the feasibility of an idea: whether it was possible to produce a car, not a micro-car, which combined advanced technologies and solutions with a price which ordinary people could afford.

The RDA remained convinced that it was impossible to achieve the sale price which Hitler wanted. In fact, when the project reached the stage of production planning, the effective costs were higher than the price set by Hitler (Nelson 1967; Nitske 1958).

At this juncture the director of BMW, Franz Popp, proposed that the design be entrusted to the German Labour Front (Deutsche Arbeitsfront, DAF), a Nazi trade union. Hitler assigned the project to the DAF, which founded the Volkswagen Company for the purpose. In 1938 Hitler attended the laying of the first stone of the factory in Wolfsburg.

In 1936 and 1937, Porsche met Ford and visited the factories at River Rouge, where he deepened his knowledge of Ford's principles, which were put into practice in the organisation of mass production around assembly lines. During the visit he contacted and hired a dozen engineers and specialist workers of German origin whom he took with him to Wolfsburg.

The Wolfsburg factory never reached full production—the mass production of cars which Hitler wanted. This was partly because, with the outbreak of war after the invasion of Poland, the factory was converted to war production, manufacturing aeroplane parts, bombs and a military version of the Volkswagen.

Mass production for the German armed forces was, however, achieved by Ford with its own German subsidiary, Ford-Werke, producing a lorry with the capacity to transport 3 tons of usable load according to a specific design of the German army. The production of Ford-Werke was not interrupted even after the invasion of France and the defeat of the Allies. Only the United States' entry into the war put an end to the commercial collaboration between the Ford Motor Company and the Third Reich (Fings et al. 2000; Overy 1975; Wallace 2003).

2.6   Prefabrication at Christoph & Unmack

Between the two wars, not many prefabrication businesses in Europe had a multinational production organisation. Exceptions were Hetzer and Christoph & Unmack, which marketed their own systems and components at an international level. Both businesses had set up a commercial system based on distribution by licence, not dissimilar to the one adopted by Hennebique; carpenters, local businesses or professionals were given franchises for prefabrication systems, with which it was possible to build a wide variety of buildings—from simple barracks, to houses for single families, blocks for several families, villas, schools, rooms, pavilions, factories—even complete hangars. Case studies, examples and floorplans were regularly published in the journals of the time (Christoph & Unmack 1935; Lewe 1920).

Christoph & Unmack began as a carpentry workshop in Copenhagen. In 1882 it began production of prefabricated systems, acquiring the patent of the Danish official Johann Gerhard Clemens Doecker for the Doeckerske Filttel; a construction system for easily transportable temporary accommodation. Doecker's patent was for a light, rapidly assembled wooden structure covered in felt. In 1883, it won the gold medal at the Hygiene Exhibition in Berlin, and the Prussian Minister of War set up two demonstration camp hospitals, built with the Doecker system, the Doecker-Bauten. Commissions began to arrive at international level, from the French war ministry and from the Russian and Austrian ones. To respond to these orders, Christoph & Unmack opened subsidiaries and factories in France at Fécamp, in Poland at Pelcowisna and in Bohemia at Friedland. These subsidiaries would meet with intermittent success.

A fundamental step for Christoph & Unmack was the opportunity to supply the Prussian Ministry of War, which in 1885 commissioned 50 barracks, either for camp hospitals or for shelters. The Minister requested that production take place in Prussia and that the supply could be extended for an indefinite period. As the Christoph family originated in Silesia, the company decided to open a new factory at Niesky, close to the existing one of their cousin John E. Christoph, which specialised in metalwork and machine tools. The two companies were to remain legally separate until 1922.

At the same time as the Doecker-Bauten production, a programme of development was begun which aimed at technical improvements to the system, primarily greater protection from heat and cold and better ventilation and transportability. It is possible to track the development of the research through the numerous patents filed by Christoph & Unmack: in 1886 for a multi-layered external roof on a support of tarred pasteboard, in 1902 for new insulators with better insulating properties to clad a corrugated cardboard panel and in 1905 for cladding panels with concealed couplings.

Orders increased rapidly. In 1895, it was estimated that production had passed a thousand barracks (Lange 1895) with an international clientele and a complete programme of light and transportable buildings: hospitals, shelters, cookhouses, living quarters, dormitories and laboratories. In the first decade of the twentieth century, the Niesky factory produced prefabricates for Argentina, Austria, Belgium, Bulgaria, Congo, Denmark, Egypt, England, France, Germany, Greece, Holland, Italy, Japan, Mexico, Romania, Russia, Serbia, Siam, South Africa, Switzerland, Turkey, the West Indies and for the Red Cross.

To respond to the need for growing production, the Niesky factory underwent several phases of reorganisation. The initial organisation was typically artisanal, with specialised workshops which made a complete part or component by hand.

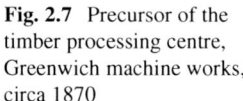

Fig. 2.7 Precursor of the timber processing centre, Greenwich machine works, circa 1870

Fig. 1.—SIX WOOD-WORKING MACHINES IN ONE.

Each craftsman began with the raw material and continued working until he had finished the piece or component. They were organised in workshops, according to the parts to be made and the necessary jobs. For example, the expert carpenters were charged with working on the wooden structure, the smiths on the metal connections, and so on.

From the nineteenth century there had been a growth of machines for processing wood, first powered by water mills, then by steam engines, and finally by electricity. The international exhibition at the Crystal Palace shows the state of the art for the era: circular saws, lathes, and machines for smoothing, pressing, boring and mortising. For working wood, mechanisation also provided greater speed, precision and repeatability than the craft techniques.

Standardised machines, acquired on the market, and specialised machines designed by Christoph & Unmack themselves, often in collaboration with the firm of John E. Christoph, began to be introduced in the processes of the Niesky factory. Initially they assisted or replaced specific craftsmen's tasks (Fig. 2.7).

At the beginning of the Second World War, the number of machine tools at Christoph & Unmack grew to more than a hundred; almost half of these were general-purpose machines which would be requisitioned to be assigned to manufacturing armaments, a wartime priority (Fokorad 1944).

From 1902, various ways of organising production were experimented with in the Niesky factory, because it seemed clear that the process of gradual substitution or addition of mechanised processes was not capable of achieving the best integration between the skills of the craftworkers and the speed and repeatability of the machine tools. A method of organisation by departments was adopted, each one specialising in a specific production. In the departments, the work was organised according to the needs for skills, be they manual or mechanical, according to a flexible organisation that combined one with the other according to needs. Changes or updates to a component could be rapidly adopted and put into production by reshaping the workflow.

Christoph & Unmack undertook standardisation of the components of the Doecker system both for production purposes, in order to optimise the use of primary materials and to reduce wood waste, a strategic problem for the wood industry, and with

> the aim of making use of a building which could be transported comfortably, assembled easily and dismantled quickly (Zorgno 1992).

The aim of standardisation is flexibility in design, the possibility of combining the components and the assembly of the greatest number of types with the least number of parts.

The Doecker patent describes a closed system of prefabrication composed of a number of elements

> which are joined together on the basis of a predetermined plan and through unique correlations (Nardi 1986).

Such a system was designed and produced by a single firm.

It is certainly a rational system, in which the number of components is reduced, using a variety of units which simplifies production, store management and the assembly and disassembly procedures of the building.

Here, standardisation has a different function from that of mass production, where the aim is joining and interchanging parts, functional to mechanised production and assembly along the production line by workers specialising in one task.

Doecker-Bauten offered a logical and flexible design system which allowed the composition of a notable variety of combinations, which could be adaptable to a variety of uses, types and climatic conditions. It is an open design system in that it is not constrained by a predefined set of layouts, although it would obviously not permit infinite variety.

The 1907 Doecker-Bauten catalogue shows 17 different types of barracks, every one of which could be personalised for different uses by varying internal fittings.

In 1907 Christoph & Unmack decided to begin the production and sale of wooden log houses, Nordische Blockhäuser, traditional in Scandinavia and Southern Germany, in which the wall beams were laid horizontally on a brick foundation, sometimes with a granite cladding, on a wider and thicker contact log.

On this contact log, blocks were laid one over another. Some logs came ready-prepared with wooden dowels and others with tightly-fitting grooves, and these were laid in turn to make tongue and groove joints. To counteract possible settlement along the vertical boards of the wall, the blocks could be fixed in place with proper dowels. The logs continued beyond the corner joint so that they fitted together at the crosspoint and stuck out at the ends (Fig. 2.8).

At the Niesky factory a floor was set up especially for the new production. Production was directed by master carpenters with extensive experience at

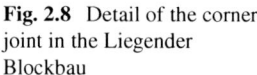

Fig. 2.8 Detail of the corner
joint in the Liegender
Blockbau

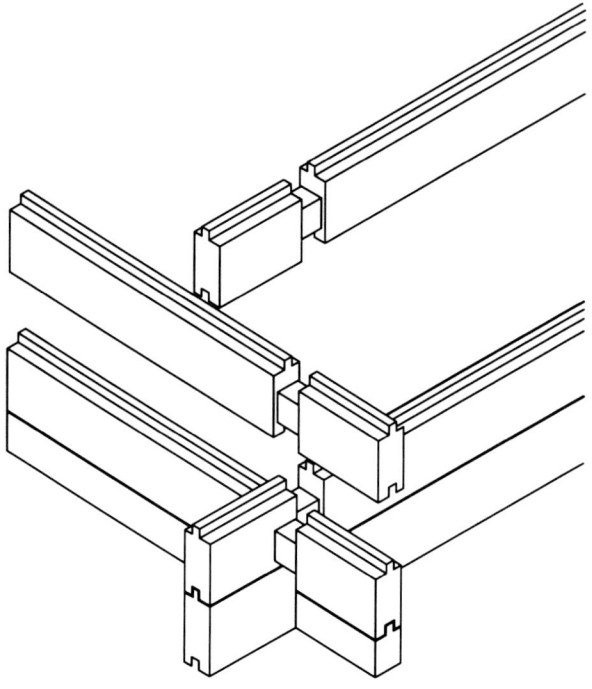

Christoph & Unmack. The new department was provided with innovative machine tools for processing solid wood in accordance with modern standards of precision and speed. The aim was to reduce the thickness of the blocks to only 2 3/4 in (7 cm), reducing the amount of solid wood needed by about 25 % compared with traditional Blockhäuser buildings. Reduced heat and acoustical insulation, due to the decreased thickness of the wood, were made up for by a multi-layered internal insulation, the result of the 1902 patent for new insulating claddings. The multi-layered system was developed further, leading to a request in 1927 for the German patent of the Swiss product Lignat. This is a large plywood board, composed of a layer of asbestos, cement, chemicals and shredded paper. At the time, the harmful nature of asbestos was not known. The panel was used extensively for cladding walls and ceilings, both because of its fire retardant, anti-condensation and insulating performance, and because of the low cost (Christoph & Unmack 1927).

Although the decreased thickness of the blocks may have diminished their overall stability, the wall was improved by the greater precision of the right angles and in the joints resulting from the mechanisation of the processes, which allowed a better use of the raw materials, lower costs for materials and assembly on site. The corners were also improved by the increased precision in the processing of the joints, so that the edges of the logs were snubbed. The purpose was to save further solid wood and to improve the statics of the walls, concentrating the loads in the central part.

Fig. 2.9 Nordische
Blockhäuser by Christoph &
Unmack: detail of the wall
and of the floor

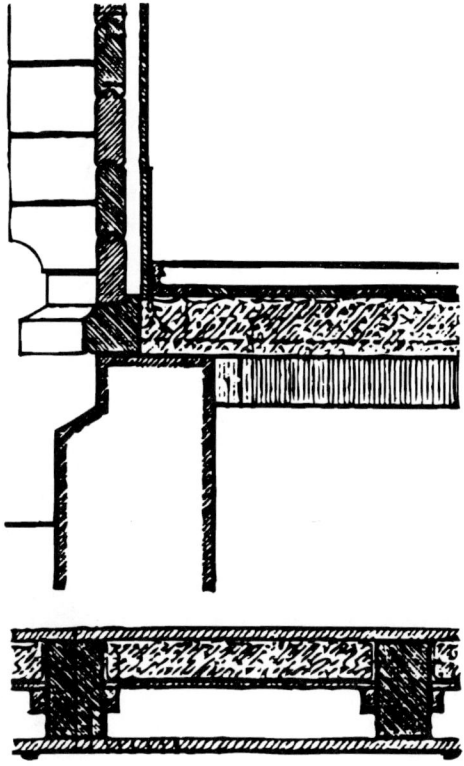

In this new system, Christoph & Unmack applied all their experience in prefabrication
and wood processing, which was particularly evident in the details, the structural
sizing, and the connections with the roof and the base (Fig. 2.9).

The end of the First World War, with the drastic fall in requests for barracks from
the war ministries, and later the Great Depression, posed difficult choices for
Christoph & Unmack. In 1926 Prof. Hans Poelzig was appointed to the Board of
Directors. The purpose was to promote collaboration between the company and the
architects, particularly with Prof. Albinmüller (pseudonym of Albin Müller), Hans
Herkommer, Joh Mundt, Hans Scharoun, Werner Schenck, Henry van de Velde,
Franz Zell and Hans Zimmermann.

In 1926 Poelzig, mentor and friend of Konrad Wachsmann, employed the latter
as head architect at Christoph & Unmack. Wachsmann describes his entry into the
Niesky factory as opening up a new world, the world of production:

> In the large factory halls I saw for the first time, like a miracle, production machines producing
> [...] prefabricated panel systems for housing, hospitals and schools, manufactured there ...
> and then shipped all over the world. In a split second I understood that mass production was
> more than a technological event. In fact, I suddenly sensed that industrialisation was the
> answer to building, and terribly important (Herbert 1984).

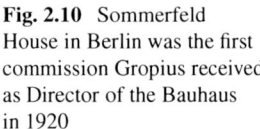

Fig. 2.10 Sommerfeld
House in Berlin was the first
commission Gropius received
as Director of the Bauhaus
in 1920

Methodologically, the collaborations between Christoph & Unmack and the various architects mentioned above had not produced the development of new construction systems. They had resulted in architecturally innovative uses of the systems which the firm had already developed, with mutual advantage: on the one side the development of new architectural languages for constructing in wood and for prefabricating, on the other the solution to technological and constructive issues raised from time to time by the designs. Wachsmann's own contribution was to perfect the system and to develop new designs. For example the "Director's house" at Niesky successfully addressed one of the crucial issues in log building: the shrinking of the wood in the direction perpendicular to the log axis.

Before processing the wood, Christoph & Unmack seasoned it in open spaces, protected from the elements, and then stabilised it in drying-kilns at constant temperature and humidity levels. However, the logs in the buildings did tend to shrink, as was also found in the traditional Blockhäuser buildings (Figs. 2.10 and 2.11).

Wachsmann calculated the entire shrinkage of the logs over time, along the axis of the log, as 4 in (10 cm) in every 9 ft 10 in (3 m). For the Blockhäuser system of prefabricated houses, Wachsmann invented a procedure to allow for

Fig. 2.11 Ernst Wasmuth Verlag AG and Birkhäuser GmbH. General view from the road of the "director's house" in Niesky

the natural shrinkage of the log so that it did not introduce tensions or deformations in the building. He designed sliding elements to match doors, windows, panelling and even guides for inserting fittings. He made detailed drawings and carried them out on the shopfloor, personally supervising the carpenters in the innovations in the production process (Wachsmann 1930; Grüning 1986). Following these innovations, a new machine tool was put into production to mill grooves, although it is not known what role Wachsmann played in the definition of the requirements or the design of this machine.

At the end of 1920, Christoph & Unmack reached the apex of architectural design: projects by well-known architects filled the journals and provided the patterns for the widespread diffusion and acceptance of their products. Catalogues were printed and distributed in runs of 5,000 copies. Later, the Great Depression, Nazism and the war intervened to change the situation radically.

2.7 General Panel Corporation

In 1941 Wachsmann joined Walter Gropius in the USA, where he became a guest at Gropius' villa in Lincoln, Massachusetts. He brought with him from the French internment camp the designs for two prefabrication systems, which would take the names of Mobilar Hangar and Packaged House (Christensen and Broadhurst 2008). At Gropius' house, Wachsmann would develop the Packaged House, starting from the initial idea of a general unitary system consisting of structural panels, externally clad to resist the elements, internally flush-panelled and thermally insulated. The panels could be freely combined to make buildings on a three-dimensional modular grid. All the main elements of the building, external and internal walls, floors, ceilings and roofs, were obtained by assembling modular variations of the panel.

The panel edges were bevelled to be connected with Y-shaped wedged connectors, which could join two, three or four panels together (Fig. 2.12).

The system designed by Wachsmann was the result of a combination of the experience acquired at Christoph & Unmack and the European heritage of research and invention in the field of prefabrication. For example, in the Hirsch Company for the prefabrication of copper houses, Gropius had already used bevelled structural panels, bolting them to a small vertical steel strut, according to the patent for the

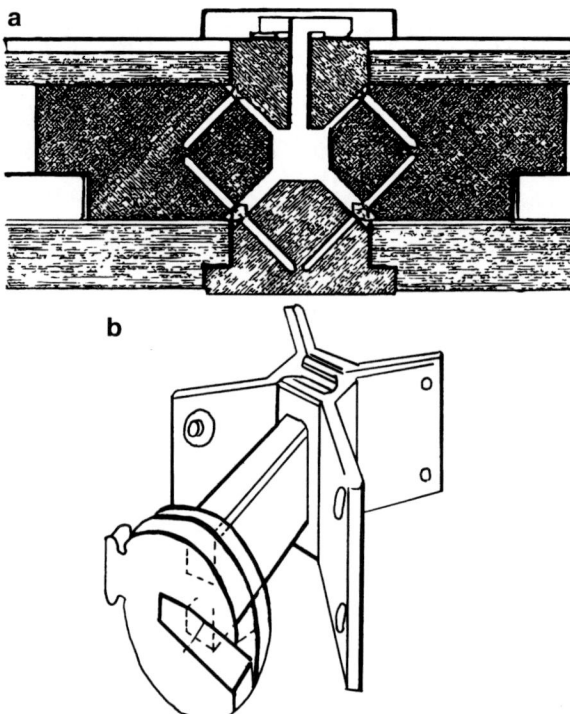

Fig. 2.12 (**a, b**) Two, three or four panels may be assembled together using Y-shaped metal connectors

"universal joint" *universalendungen*, granted to the architects Friedrich Förster and Robert Krafft (Von Borries and Fischer 2009).

Wachsmann worked with untiring dedication to perfect the system. He and Gropius filed the first patent request for a prefabricated building in 1942.[1] The patent request states:

> This invention relates to buildings intended for various purposes, such as dwellings, camps, barracks, and others.
>
> It is the chief object of the invention to devise a building structure which can be assembled exclusively, or substantially so, from standard units or sections, each consisting fundamentally of a duplicate of the other, so that they can all be manufactured completely in a factory equipped with machinery for producing them efficiently and accurately and thus eliminating much of the labor customarily involved in manufacturing parts at the site of the building. In other words, the invention aims to transfer most of the labor involved in the construction of a building from the site of the building itself to a factory and to make the erection of the building primarily one of assembly.
>
> To this end it is a further object of the invention to devise a construction which will eliminate practically all of the necessity for using nails, screws, hooks, and similar fastenings during such assembly, but to provide a more satisfactory means for securing the sections together and to make the sections so standardized that with only minor exceptions, any frame section can be interchanged with any other.
>
> [...] A novel form of wedge connector or joint is provided for locking the building units in this relationship and the parts of which this joint is composed are illustrated individually in Figs. 8, 9 and 10. They consist of a plate connector or main-stay 10 having two parallel slots near its opposite ends, a cross piece 12 one portion of which is offset relatively to the other, both being slotted, and a wedge 13.
>
> [...] Such a wedge connector can be and preferably is used throughout the structure. Important advantages of it are that it permits the union of the parts to each other rapidly without the use of tools other than a hammer, does not require skilled labor, and that it lies entirely within the frame pieces which it connects. Inasmuch as the four arms of this joint body radiate from a common axis, and the diagonal edge faces of all the pieces connected together by it intersect at said axis, and this axis is at the center of the joint, the action of the wedges as they are driven into place forces all of these wood pieces radially toward said axis until such movement is arrested by the meeting of the obliquely disposed or bevelled surfaces of those pieces.
>
> [...] It is the intent of this invention that, so far as practical, all of the manufacturing operations on each panel shall be completed in a factory so that the finished sections, each including its own frame, may be shipped to the site of the building ready for assembly into a complete structure. These standard parts can be assembled into various designs of building.

In 1942 Wachsmann and Gropius founded the General Panel Corporation, funded by the investment bank Charles Allen & Co. and based in New York, with the purpose of designing, producing and marketing the patented system.

Wachsmann continued to refine the system, even at the expense of its manufacture and commercialisation, the main purpose of the investment by Charles Allen & Co. Wachsmann's attention was focused on the connector component: he designed version after version before having a new idea, for which he filed a new patent application in 1945 (Figs. 2.13 and 2.14).[2]

---

[1] United States Patent Office Patent n. 2.355.192.
[2] United States Patent Office Patent n. 2.421.305.

This invention aims to improve the wedge connectors used in said building in such a manner that all of these connector elements can be installed in the panels, or other building units, at the factory, thus simplifying the assembling operation at the site of the building and reducing the labor of assembly.

In addition to the advantages of this construction above referred to, it also permits the use of either smooth, grooved, or other special siding and lining materials, and the sealing strip which was required in the earlier construction has been eliminated.

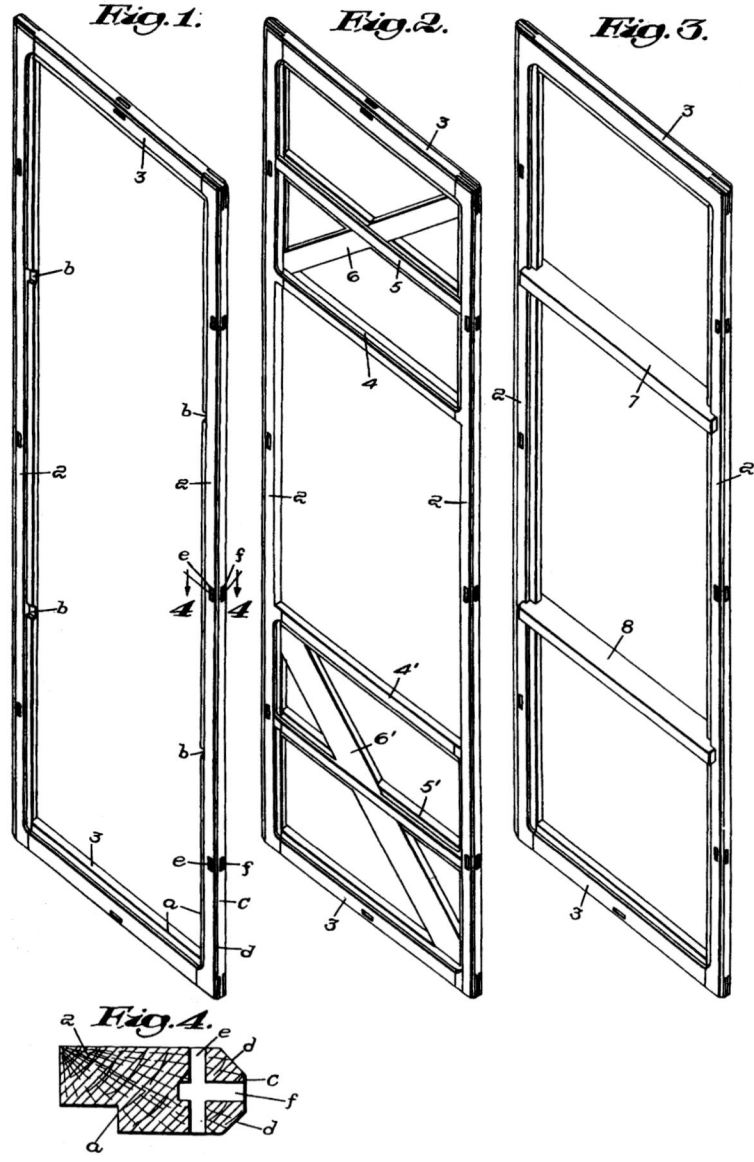

Fig. 2.13 United States Patent Office Patent no. 2,355,192, filed by Wachsmann and Gropius in 1942

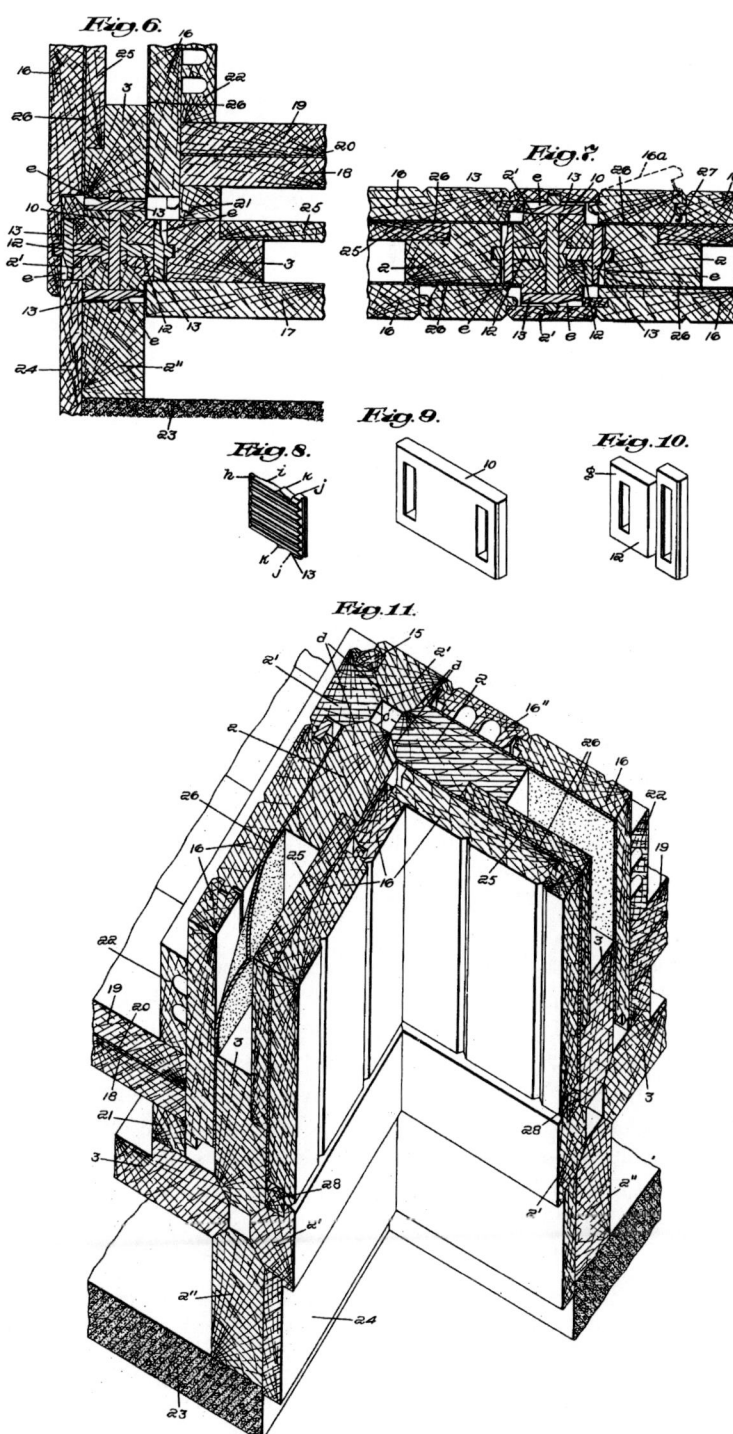

Fig. 2.13 (continued)

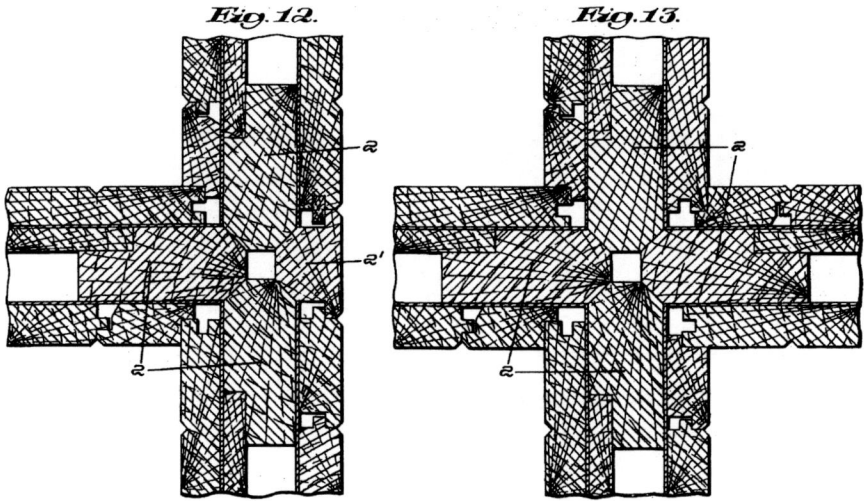

Fig. 2.13 (continued)

It was a new type of connector, made with interlocking metal plates, mounted in the panel edge. This connector was made starting with flat parts, stronger and simpler to produce than the initial design, based on an articulated three-dimensional Y-shaped form which had to be fixed using wedges and then screws (Fig. 2.15).

Packaged House set up a three-dimensional *design system*, based on a standard grid 40 in (102 cm) square, on which the panels were combined to build walls, partitions and roofs. It was a flexible system to the extent that Wachsmann deliberately chose never to make a definitive design of a building that used it, as such a reference point or pattern would have restricted the creative possibilities of an open method (Fig. 2.16).

The Packaged House design system was however proprietary, self-sufficient and expressed in a number of purpose-made parts. Because of the proprietary panels and joints, it excluded the possibility of integrating the numerous industrialised components sold by other companies around on the market at the time, including prefabricated windows, doors, roofs or foundations. It was a closed system, intended to integrate a certain number of coherently designed components that were manufactured to conform to a unitary production system. As a result, the system was difficult to reconcile with existing standards, such as the USA plywood sheet standard width of 48 in (122 cm) (Herbert 1984).

Packaged House was designedly different from the systems based on

> elements which can be assembled in several ways, since the connection, the organisation, and the coordination among the parts are guaranteed, allowing freedom of design (Nardi 1976).

In the United States, the paradigm of an open system is the balloon frame system, promoted by the economic advantage of timber and nails manufactured and sold on an industrial scale. The system frames were available on the market as boards

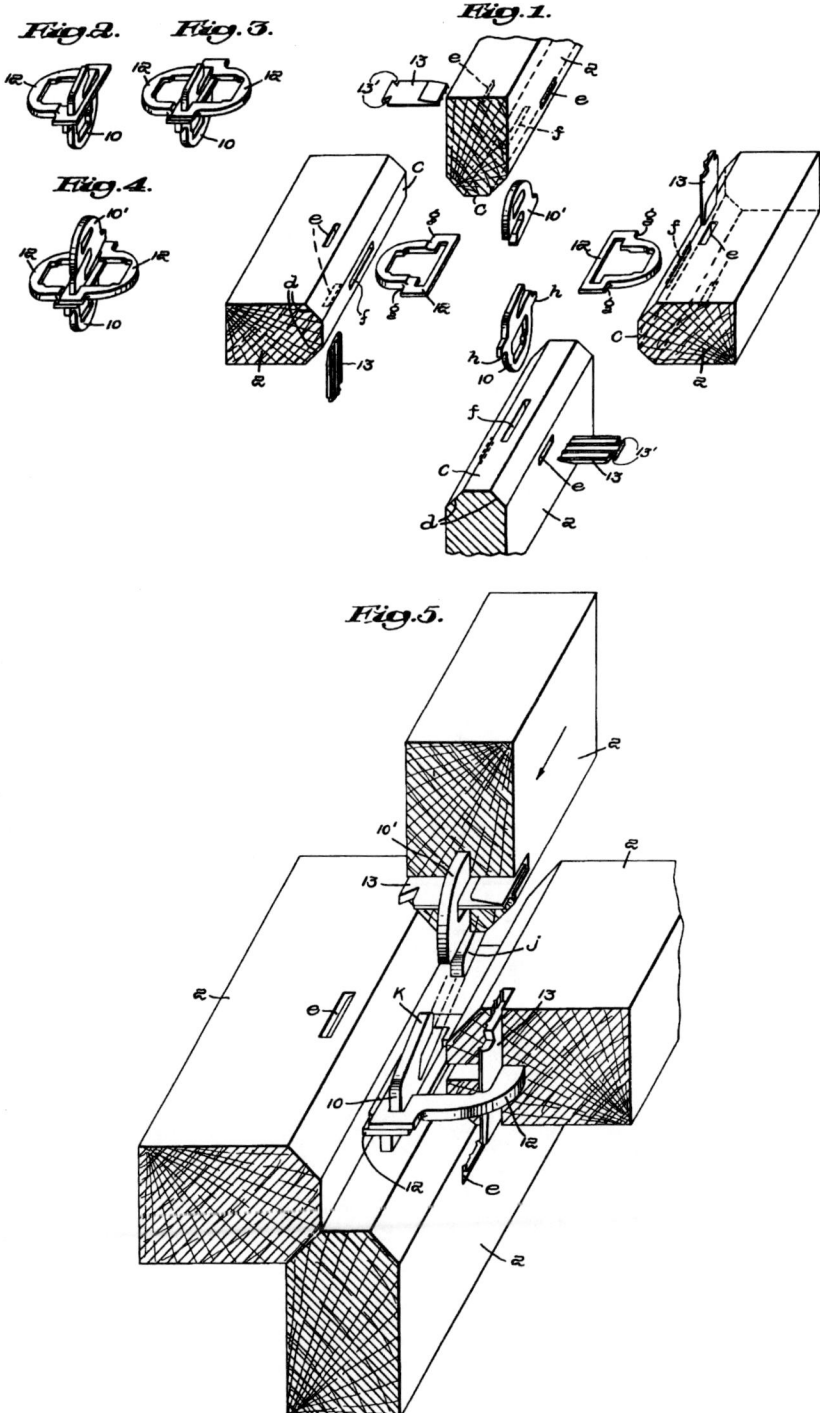

Fig. 2.14 United States Patent Office Patent no. 2,421,305, filed by Wachsmann and Gropius in 1945

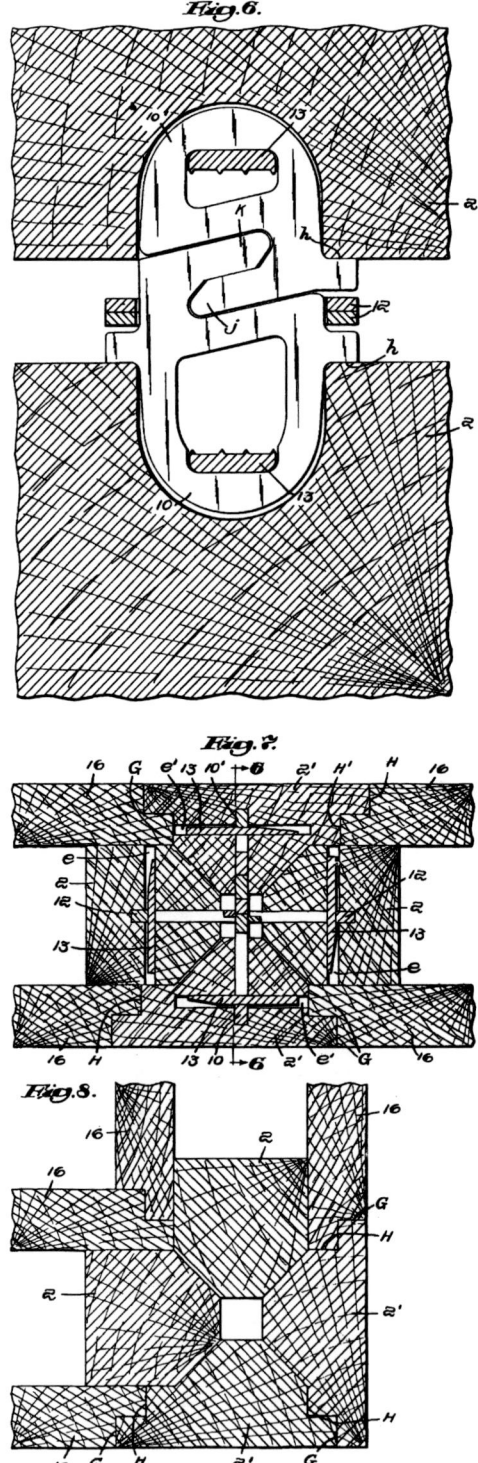

Fig. 2.14 (continued)

Fig. 2.15 The new patent connector made with flat parts: more robust and simpler to manufacture

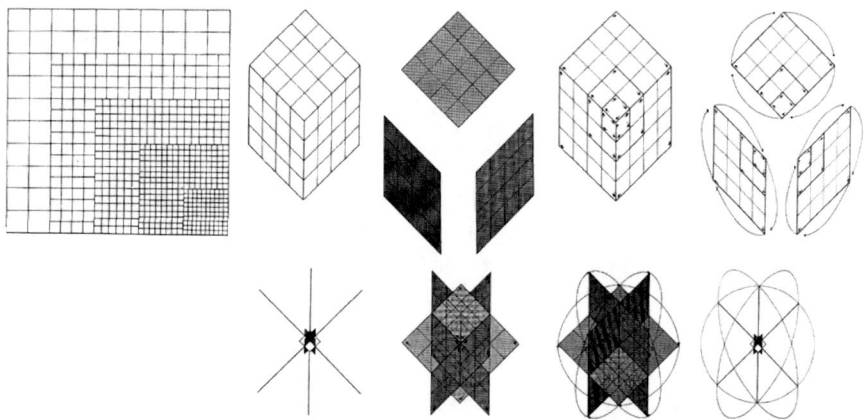

Fig. 2.16 Wachsmann diagrams for modular coordination in two-dimensional space, three-dimensional space and in the time dimension for assembly

usually in the dimensions of 2 by 4, 2 by 6, 2 by 8 and 2 by 10 in, and 4 ft long (5 by 10, 5 by 15, 5 by 20 and 5 by 25.4 cm, and 1,2 m long). These dimensions were the result of a rule of thumb, mainly among sawyers and carpenters. This agreement was at first not even written down (Dupire et al. 1985), but was rather the outcome of a gradual consensus on the units of measurement, inches and feet;

the economies of scale which could be made by the sawyers, in terms of work and waste; transportation, on lorries and by rail; and handling and crafting on the construction site.[3]

General Panel made various complete prototypes of houses and presented them to the government, the military and the media, who were enthusiastic about them. But production did not begin, as the war was ending and with it the huge sales opportunities for prefabricated buildings which it had created.

The post-war period still offered good prospects to the prefabrication industry. In 1946, President Truman set up the Office of Housing Expediter in order to manage the return home of war veterans and workers in the war industry, giving rise to the provision of 1.2 million residential units for that year and the expectation of bigger contracts for the following year.

Celotex Corporation offered General Panel a commercial agreement with the aim of producing and distributing the Packaged House system. Celotex produced multilayered plywood panels clad in asbestos, foreseeing that the panel system patented by Wachsmann would be an outlet for their products. To this end, in 1946 the General Panel Corporation was set up in Burbank, where they acquired the disused Lockheed plant in which the P-80 Shooting Star plane had been produced during the war. The agreement specified that the New York office would manage design and development, whilst the one in Burbank would manage production and sales. Wachsmann was the Chairman of the New York branch, whilst Carl Dahlberg, son of the Chairman of Celotex, was Chairman of the Burbank branch.

The plans were to endow the Burbank factory

> with the most sophisticated, complex, and expensive production plant. Together with some remodeling of the Lockheed factory, design of the plant layout had been going on intensively throughout the second half of 1946. Most of the work was conceived, designed, and detailed by Wachsmann and his New York office (Herbert 1984).

In 1947, the plant was ready for the industrial manufacture of the Packaged House system. At the time, it was one of the most advanced, integrated and automated plant of the sector. Only the factory of Lustron Homes near Columbus, fitted out at the beginning of the 1950s, could rival it (Fig. 2.17).

The plant conformed perfectly to Wachsmann's philosophy which permeated the design of both the prefabricated and the manufacturing systems:

> The standardizing of our basic product makes them highly adapted to modern machine needs. We have equipped our plant with specially designed machinery of a high-speed, multiple-operation character; radio frequency installations which cure glue bonds in a matter of seconds; and advanced quality control devices which assure steady and reliable quality in the product of every work station. These processes are repetitive and highly adapted to straight-line production, since the universal joint means that all panels, whether for floors, walls, partitions, ceilings or roofs, are structurally similar from the standpoint of production (Herbert 1984).

The production line progressed in the factory exactly as designed: the redwood timber, supplied according to precise specifications regarding moisture and thickness,

---

[3] See Sect. 2.8.

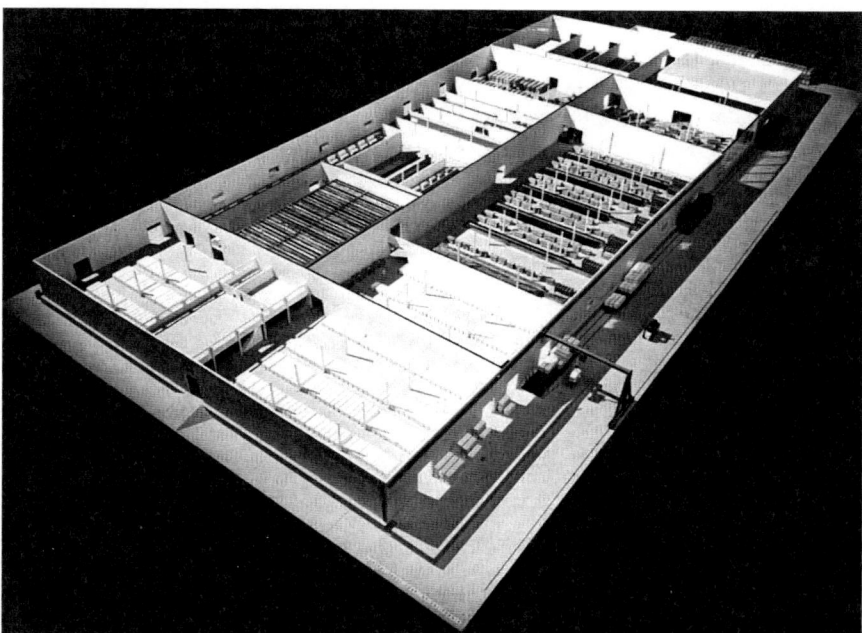

Fig. 2.17 Isometric view of the General Panel Corporation plant in Burbank. "A production line. Symbol of the concept of automation. A certain number of machine tools, mechanically and chronologically synchronised and servo-driven, shape the materials into the finished product" (From: Wachsmann (1961))

was loaded by conveyor belts and moved to a series of automatic hoppers. The panel frame was produced and assembled along the line, being first planed with high speed mechanical planes and then passing through double end tenoners, jointmakers and multi-slot mortising machines. All the components, the panel frame, the structural elements and the filler strips, passed along the line for the insertion of the cadmium-plated die-cast steel connectors. The steel panel frames passed through glue-spraying machines.

Where required, an insulation layer was added. The sheets of Douglas fir plywood arrived along conveyor belts to be glued to the frames in only 5 s with special high-frequency machines. An electronic press consolidated the entire stressed-skin panel in a single operation.

The production line ended with painting, where each part was sprayed with primer and undercoat in preparation for the final coat of paint on the construction site after assembly of the panels. The piles of panels and structural parts were stocked in lots, one for each order. The lots moved towards the packing area, to be loaded onto lorries and sent to construction sites. Each order was complete with every part and component needed to assemble the finished building in its place, including connectors, insulation, frames, doors, windows and gutters, with the concealed wiring.

The story of General Panel Corporation does not have a happy ending. In 1948, when the factory was working at full stretch, management of the emergency by the Office of Housing Expediter was ended. The company offered the panels and the system on the market, which was mainly for single family houses. To assist buyers to choose for themselves, General Panel printed catalogues of finished houses with prices for comparison. In these catalogues, the potential of the design system were sacrificed to the end result: the American family was interested in the finished house.

The finish of the Packaged House was above average and so, therefore, were their prices. General Panel Corporation's costs depended on the costs of the investment needed to build a cutting-edge production line, as well as on operating costs, employees' wages, raw materials, maintenance and shipping.

The panels were made of expensive types of wood: redwood and Douglas fir. Moreover, the standard unit of the system, 40 in, did not match the USA standard width of sheets of plywood, 48 in, which meant that each panel produced 8 in of waste material. The structural parts, including columns and windowsills, were moulded from blocks of high quality solid wood, a costly process. The electrical fittings were concealed within the panels, which apart from being a costly proce-dure led to disputes with the trade union of the craft workers involved. All these factors combined to raise the price of Packaged Houses above the norm both for those made using the widespread *platform frame* system and for those of the more than 70 manufacturers of prefabricated homes in the United States (Kelly 1951, Fig. 2.18).

Lewis Mumford's 1930 remarks on prefabrication proved to be correct, particu-larly his opinions on the influence of various costs on the final price, not least the need to organise a distribution system with credit facilities suitable for real estate, albeit built with prefabrication. The American banks tended to apply the same con-ditions and charges for prefabricated houses as they did for mobile homes, both classed as nondurable goods.

Wachsmann reviewed his own experience in prefabrication on several occasions. In 1961, in his book *The Turning Point of Building: Structure and Design*, he affirmed the role played by industrialisation:

> In discarding many of our old ideas about building, we have reached a turning point. The decisions about what constitutes the formative energies of the age have been made and the principles that will guide the developing forward movement are now apparent (Wachsmann 1961).

The analysis of the reasons for General Panel Corporation's lack of success shows that, besides incidental factors which might have been remedied such as expensive raw materials, or those which might have been improved such as the wastage of materials during processing, there were also structural problems which not even the cutting-edge Burbank factory had been able to solve. Despite the high level of automation of the manufacturing process based on numerous machine tools, the factory needed around 500 workers along the production line, particularly in the assembly stages.

Fig. 2.18 Sheets of Douglas fir plywood are glued into the frames in only 5 s with special high-frequency machines. An electronic press consolidates the entire stressed skin panel in a single operation

After the introduction of assembly lines at the Ford factory, an article by the managing director of Smith Corporation asked, probably as a challenge:

Can automobile frames be built without men? [...] We set out to build automobile frames, without men. We wanted to do this on a scale far beyond that necessary to meet the immediate requirements of the automobile industry (Smith 1929).

After the experience at General Panel, Wachsmann's research continued coherently with his philosophy: that systems of design and production should be conceived as a single, integrated, organic unit.

In this, the assembly, whether done in the factory or completed on-site, was at odds with of the aim of an organically unified process. Towards the end of the 1960s, Wachsmann addressed his research to further automation of the manufacturing process, trying to solve the problem of assembly. In his laboratory at the University of California in Los Angeles, where he taught from 1964, he developed the design for a "manipulator", which was able to move, to line up and to assemble generic elements or components in three dimensions. The goal of the research was

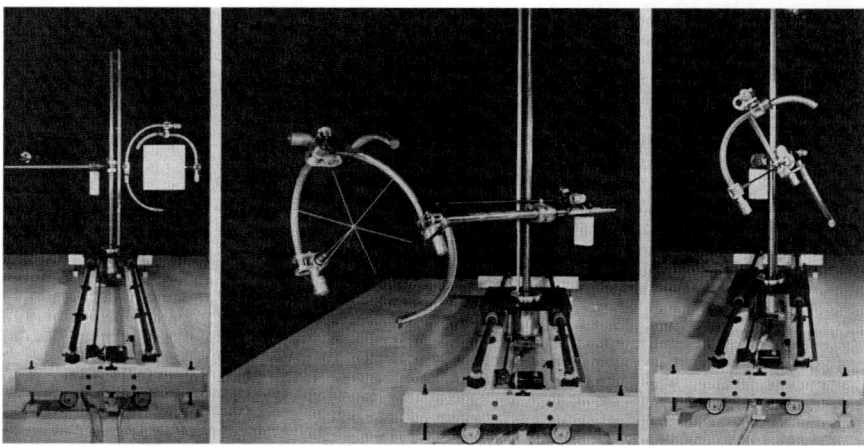

Fig. 2.19 Working prototype of anthropomorphic robot with seven degrees of freedom, made by Wachsmann in his Laboratory at the University of California towards the end of the 1960s

a substantial step forward, both with regards to the machinery currently used in the production lines, like conveyor belts and trolleys, and to the cranes normally used on construction sites. So that the manipulator could grasp and arrange pieces in any position, Wachsmann broke down the space of movement in the time-dimension, a procedure in which he already had solid experience since his representation and decomposition of the modular space with diagrams (Figs. 2.16 and 2.19).

On the basis of these studies, he defined the requirements for a manipulator with seven degrees of freedom (DOF) of movement, permitting it to grasp or install a piece in any position. This was followed by designs for several machines able to satisfy these requirements.

The solution was found in the emerging science of robotics. Towards the end of the 1950s, George Devol and Joseph Engelberger invented a mechanical arm and associated it with the concept of "universal automation" with the aim of giving further impulse to the emerging sector of computerised control. The prototype evolved into the Unimate, probably the first robot to be manufactured and marketed.

The Unimate had three or four degrees of freedom of movement; Wachsmann's prototype had seven. Wachsmann's solution was obtained through translation in Cartesian space (three degrees of freedom), three Euler angle rotations (3°) and finally one polar rotation (1°) (Bock 2009). The working prototype was named the Location Orientation Manipulator (LOM), as it could solve any task of movement and assembly it might be assigned; the seven degrees of freedom allowed a flexibility of movement equivalent to that of the human arm. However, it was not until the first years of the twenty-first century that industrial robots with seven effective degrees of freedom began to be available on the market.

2.8   The Balloon Frame System

The history of the balloon frame prefabrication system is not the history of a firm nor of a designer. Even its origins are disputed.

Although it is known to have appeared in the Chicago area at the beginning of the 1830s (Sprague 1981), it is not certain who invented it. Nineteenth century historians in particular, on the strength of the idea of inventions and patents, tried to name the source of this popular construction system, which was attributed by some to George Washington Snow and by others to Augustine Taylor. In those years, many carpenters were erecting several buildings in Chicago with this new system. Even the source of the name is uncertain, although the negative meaning linked to its lightness—as light as a balloon that the wind could blow away—seems to be the most commonly held view (Cavanagh 1997).

A whole generation had to pass before the system was developed to its full capacity and began to spread rapidly and widely with the new settlements:

> The venture was risky. It was also progressive because it focused on good profit for the time, money, and labor invested. The adoption of balloon frame construction happened in the same atmosphere of agricultural entrepreneurship. When carpenters and farmers understood the internal logic of the balloon frame they were thinking in the new categories of rational, efficient use of time, energy, and materials. Balloon frame construction was an integral part of the growth of a market economy that became the basis for the nation's wealth. Building balloon frame structures meant abandoning the old ways for the new and moving ahead into a productive future (Peterson 2008, p. xi).

The adoption of the new construction system followed the settlements of the pioneers. Peterson (2008) estimates that 90 % of the farmhouses in the Midwest were built with the balloon frame system.

In 1846, Solon Robinson, a farmer and pioneer, wrote one of the first descriptions of the system in an article for American Agriculturalist, *A Cheap Farm House*:

> Intended for the new settler, and to be built on the balloon plan, which has not a single tenon or mortise in the frame, except the sills; all the upright timber being very light, and held together by nails, it being sheeted under the clapboards is very stiff, and just as good and far cheaper than ordinary frames (Robinson 1846). (Fig. 2.4)

The system underwent a series of improvements through a widespread process of trial and error. One example was the many attempts to improve the insulation of the walls:

> Considerable discussion about the proper way to insulate the walls of a balloon frame house added to the controversies about the system (Peterson 2008, p. 23).

Only towards the end of the 1870s did the system, its component parts and the construction procedures begin to come together and to produce standardised terminology and techniques.

Balloon framing, as a construction system, still widely adopted especially in its later developments such as the platform frame (Fig. 2.5), is based on a number of semi-finished components for building the structure (the cladding of the façade,

Fig. 2.20   Hydraulic sawmill in America at the beginning of the nineteenth century

the openings and all the rest) together with methods of fastening them together. The semi-finished components were made by a variety of different manufacturers. Putting them together depended on the skill of the workers, who for most of the new buildings were expert carpenters. But self-building was not unusual, mainly for house extensions in order to accommodate growing families (Fig. 2.20).

Standardisation of the construction elements was however already substantially complete by that time. It resulted from the wide availability of industrially manufactured timber and nails and from a process of defining conventions (at first not even written down) among the experts of the sector, who were influenced by the constraints posed by the means of production and transport.

Dupire (1985) called the process which developed "mainly within the United States timber industry", *American chain*. Its most noteworthy characteristics were: industrial production of simple semifinished structural elements on a vast scale through a variety of firms which compete commercially;

> the installation draws on the skills of qualified workers, able to take the responsibility of adapting the products to the requirements of the specific building (Dupire 1985)

and a service network aimed at supporting the integration of the semifinished component from the design via installation and subsequent maintenance.

This chain originated in the United States and spread to Europe and throughout the world, because of its proven flexibility in the way it could adapt to the requirements specific to the building industry. In 1956, Gropius signalled and theorised about the emergence of the process:

> organisation must therefore aim first of all at standardising and mass-producing not entire houses, but only their component parts which can then be assembled into various types of houses, in the same way as in modern machine design certain internationally standardised parts are interchangeably used for different machines. The production policy would provide for carrying in stock all individual parts necessary for the construction of houses of various types and sizes, to be ordered to the building site as required from various specialised factories. At the same time field-tested assembly plans for houses of different layout and appearance will be available to the public. Since all the standardised machine-made parts will fit together accurately, house erection at the site on the basis of precise assembly plans can be performed rapidly and with a minimum of labour, partly with unskilled workers, and under any conditions of weather or season (Gropius 1956).

The industrialisation of timber processing and the manufacture of nails turned out to be crucial for the development and growth of the American chain.

In the United States of 1920, the per capita consumption of timber was four times as high as that of England which, as a consequence of the scarcity of this material, had experienced a grave crisis held to be crucial in the adoption of coal as the primary energy source (Rosenberg 1976). Notable efforts were made in the industrialisation of the sector as it spread throughout the USA; sawmills were erected in the wake of the new settlements. At the beginning of the nineteenth century the most widespread technology was the water mill for powering saws. In a country of vast forest resources, thick blades were adopted, giving priority to sawing speed at the expense of saving material, "wasting as much as three-eighths of an inch of timber per cut (10 mm)" (Thomson 2009), ten times thicker than a similar process in England.

It was Robert Hoe and Henry Disston who, in about 1870, hired expert British saw makers to introduce innovative saws into the USA that were more accurate and productive (Disston 1915).

Mechanized saws spread quickly and widely through the United States, meeting ever increasing demands. Timber consumption in boards per capita went from 59 ft (18 m) in 1799 to 65 ft 7 in (20 m) in 1829, and then quadrupled to 262 ft 5 in (80 m) in 1859 (Thomson 2009). It was industrial manufacturing that standardised the balloon frame system and made it profitable.

> The more efficient and economical band saw came into use in the 1880s, and by the 1890s only skilled operators milled lumber. Steam-driven saws and carriages cut the logs into large sections. A conveyor sent these slabs to another saw to be cut into dimension lumber, trimmed to standard length, and sorted according to size (Peterson 2008).

The rapid development of timber processing industries and nail manufacturing industries and the parallel growth of the balloon frame cannot be explained only in terms of the relationship between supply and demand. The introduction of a large number of industries able to manufacture products—nails and logs—that were

Fig. 2.21 (**a**, **b**) LaBelle Works factory in Wheeling, Ohio, for the industrial production of nails using machines for cutting and manufacturing the metal wire

substantially similar in quality and price through the wide extent of the United States, within the time-span of a decade or two, was made possible by the development of manufacturing systems based on widely used machine tools. If we study in detail the development of the manufacturing process, whether in the timber industry or the nail industry, we can see that both depended on a driving power (chronologically first water, then steam and finally electricity) and on the introduction of manufacturing processes supported by specific machine tools, respectively the continuous band saw and the mechanically powered trip hammer (Fig. 2.21).

The wide availability of these machine tools and their continuous improvements and developments were due to the existence of a specialised industry. This highly qualified industry has specific traits: it requires specialist knowledge in the field of precision mechanics in order to produce machinery as well as further knowledge of the manufacturing processes, in which those machines are used. The role of the machine tool industry should not be judged merely on the basis of the number of parts produced, the number of employees or its turnover. It has a vital role in the development of technology and in the dissemination of the processes, without which the pace of industrial development would have been much slower.

The mechanisation of the sector was due to the growth of a new system for the transmission of know-how, whereof machine tools were an integral part. In the USA, this system was able to establish links between businesses using the new technologies, businesses manufacturing these technologies, the market, research institutions, educational institutions and government agencies. Government agencies acted as promoters of the development by means of commissions, as funders of research and as guarantors through the patenting process.

The machine tool industry made a fundamental contribution to the development and industrialisation of timber processing from 1830 onwards. An important manufacturer in the field of machine tools for timber, J.A. Fay, offered 56 different machines in his 1856 catalogue. The Census of Manufacturing of 1860 counted 23 firms specialising in machine tools for the timber sector (Thomson 2009).

Rosenberg defines the way this system developed as *technological convergence.*

This convergence exists throughout the machinery and metal-using sectors of an industrial economy. Throughout these sectors there are common processes, initially in the refining and smelting of metal ores, subsequently in foundry work whereby the refined metals are cast into preliminary shapes, and then in the various machining processes through which the component metal parts are converted into final form preparatory to their assembly as a finished product. It is with the machinery stages, of course, that we are primarily concerned here.

The use of machinery in the cutting of metal into precise shapes involves, to begin with, a relatively small number of operations (and therefore machine types): turning, boring, drilling, milling, planing, grinding, polishing, etc. Moreover, all machines performing such operations confront a similar collection of technical problems, dealing with such matters as power transmission (gearing, belting, shafting), control devices, feed mechanisms, friction reduction, and a broad array of problems connected with the properties of metals (such as ability to withstand stresses and heat resistance). It is because these processes and problems became common to the production of a wide range of disparate commodities that industries which were apparently unrelated from the point of view of the nature and uses of the final product became very closely related (technologically convergent) on a technological basis [...].

This technological convergence had very important consequences for both (1) the development of new techniques and (2) their diffusion, once developed (Rosenberg 1963).

An alternative way of stating this is to say that, in the area of machine technology, firms became increasingly specialised by process rather than product (Rosenberg 1970c).

References

Blitzer WF (1938) Case study of an entrepreneur: Foster Gunnison. The Architectural Forum 69(11). In: Kelly B (1951) The prefabrication of houses. A study of the Albert Farwell Bemis Foundation of the prefabrication industry in the United States. MIT Press/Wiley, Cambridge, MA/New York

Bock T (2009) Turning points in construction. In: 26th ISARC 2009 International symposium on automation and robotics in construction, International Association for Automation and Robotics in Construction, Austin

Cavanagh T (1997) Balloon houses. The original aspects of conventional wood-frame construction re-examined. J Archit Educ 51(1):5–15

Christensen P, Broadhurst R (2008) Home delivery: fabricating the modern dwelling. Museum of Modern Art, New York

Christoph & Unmack AG (1927) Bericht vom 1. November 1926 bis 31. Betriebsarchiv, Niesky

Christoph & Unmack AG (1935) 1835–1935. 100 Jahre Christoph & Unmarck. Oberlausitz, Niesky

Crespi R (1967) Modulo e progetto. Note in margine al rapporto Coordinazione dimensionale nell'edilizia ONU, Comitato per l'abitazione, la costruzione e la pianificazione. Dedalo libri, Bari

Disston H & Sons (1915) The saw in history. Keystone saw, tool, steel and file works, Philadelphia. Cited in Thomson R (2009) Structures of change in the mechanical age. Technological innovation in the United States 1790–1865. The Johns Hopkins University Press, Baltimore

Dupire A et al (1985) L'architettura e la complessità del costruire, Milano, CLUP. French edition: Dupire A et al (1981) Deux Essais sur la Construction, Mardaga, Liège

Fings K et al (2000) Working for the enemy: Ford, general motors, and forced labor in Germany during the second world war. Berghan Books, New York

Fokorad (1944) Technische Unterlagen für das DWH-Behelfsheim, Typ DWH 1002, Der Reichsbeauftragte für den Holzbau

Gartman D (2009) From autos to architecture: Fordism and architectural aesthetics. Princeton Architectural Press, New York

Giedion S (1948a) Standardisation and interchangeability. In: Giedion S (ed) Mechanization takes command: a contribution to anonymous history. Oxford University Press, New York

Giedion S (1948b) The assembly line and scientific management. In: Giedion S (ed) Mechanization takes command: a contribution to anonymous history. Oxford University Press, New York

Gropius W (1956) Scope of total architecture. George Allen & Unwin, London

Grüning M (1986) Der Wachsmann-report: Auskünfte eines Architekten. Verlag der Nation, Berlin

Heidegger M (1977) Building dwelling thinking. In: Krell DF (ed) Basic writings: from "Being and time" (1927) to the "Task of thinking" (1964). Harper & Row, New York

Herbert G (1984) The dream of the factory-made house: Walter Gropius and Konrad Wachsmann. MIT Press, Cambridge, MA

Hill F, Wilkins M (1964) American business abroad: Ford on six continents. Wayne State University Press, Detroit

Hooker C (1997) Life in the shadows of the Crystal Palace 1910–1927: Ford workers in the model T era. Bowling Green State University Popular Press, Bowling Green

Hounshell DA (1987) From the American system to mass production, 1800–1932: the development of manufacturing technology in the United States. Johns Hopkins University Press, Baltimore

Jefferson T, Holmes J (2002) Thomas Jefferson: a chronology of his thoughts. Rowman & Littlefield, Lanham

Kelly B (1951) The prefabrication of houses: a study of the Albert Farwell Bemis Foundation of the prefabrication industry in the United States. MIT Press/Wiley, Cambridge, MA/New York

Lange W (1895) Der Barackenbau, mit besonderer Berücksichtigung der Wohn – und Epidemie-Baracken. Baumgärtner, Lipsia

Lewe Y (1920) Die Berechnung des geschlitzten Ringdübels, System Tuchscherer. Holzbau, Beilage der Deutschen Bauzeitung 20

McLaughlin GC (1956) Eli Whitney and the birth of American technology. Little, Brown & Co., Boston and Toronto

Milward AS (1977) War, economy and society, 1939–1945. Allen Lane, London

Mumford L (1934) Technics and civilization. Harcourt, Brace and Co., New York

Mumford L (1945) City development; studies in disintegration and renewal. Harcourt, Brace and Co., New York

Nardi G (1976) Progettazione architettonica per sistemi e componenti: contributi didattici alla progettazione per l'edilizia industrializzata. Franco Angeli, Milano

Nardi G (1986) Le nuove radici antiche. Saggio sulla questione delle tecniche esecutive in architettura. Franco Angeli, Milano

Nelson WH (1967) Small wonder: the amazing story of the Volkswagen. Little, Brown & Co., Boston/Toronto

Nitske WR (1958) The amazing Porsche and Volkswagen story. Comet Press Books, New York

Olmo C (ed) (1994) Il Lingotto, 1915–1939. L'architettura, l'immagine, il lavoro. Umberto Allemandi, Torino

Olmo C, Comba M, di Pralormo MB (2003) Le metafore e il cantiere. Lingotto 1982–2003. Umberto Allemandi, Torino

Overy R (1975) Cars, roads, and economic recovery in Germany 1932–38. Econ Hist Rev New Ser 28(3):466–483

Peterson FW (2008) Homes in the heartland: balloon frame farmhouses of the upper midwest, 1850–920. University of Minnesota Press, Minneapolis

Raff DMG (1991) Ford welfare capitalism in its economic context. In: Sanford MJ (ed) Masters to managers: historical and comparative perspectives on employers. Columbia University Press, New York

Robinson S (1846) A cheap farm-house. Am Agric 5(2):57

Rosenberg N (1963) Technological change in the machine tool industry, 1840–1910. J Econ Hist 23(4):423

Rosenberg N (1970a) Economic development and the transfer of technology. Some historical perspectives. Technol Cult 11(4):557

Rosenberg N (1970b) Economic development and the transfer of technology. Some historical perspectives. Technol Cult 11(4):558

Rosenberg N (1970c) Economic development and the transfer of technology: some historical perspectives. Technol Cult 11(4):553

Rosenberg N (1976) Perspectives on technology. Cambridge University Press, Cambridge

Rosenman D (1942) Housing to speed production. Archit Rec 91(4):42–46

Smith LR (1929) We build a plant to run without men. The Magazine of Business, February. Cited in: Braham W, Hale JA (eds) (2007) Rethinking technology: a reader in architectural theory. Routledge, New York

Smith MR (ed) (1985) Military enterprise and technological change: perspectives on the American experience. MIT Press, Cambridge, MA/London

Sprague PE (1981) The origin of balloon framing. J Soc Archit Hist 40(4):311–319

Talanti AM (1975) Storia dell'industrializzazione edilizia in Italia 1945–1974. AIP, Milano

Thomson R (2009) Structures of change in the mechanical age: technological innovation in the United States 1790–1865. The Johns Hopkins University Press, Baltimore

Von Borries F, Fischer JU (2009) Heimatcontainer. Deutsche Fertighäuser in Israel. Edition Suhrkamp, Frankfurt

Wachsmann K (1930) Holzhausbau. Technik und Gestaltung. Ernst Wasmuth Verlag AG, Berlin. New edition: Wachsmann K (1995) Holzhausbau. Technik und Gestaltung. Birkhäuser, Basel/Berlin/Boston

Wachsmann K (1961) Turning point of building: structure and design. Reinhold Pub. Corp, New York

Wallace M (2003) The American axis: Henry Ford, Charles Lindbergh, and the rise of the Third Reich. St. Martin's, New York

Woodbury RS (1972) Studies in the history of machine tools. MIT Press, Cambridge, MA

Zorgno AM (ed) (1992) Konrad Wachsmann. Holzhausbau costruzioni in legno, tecnica e forma. Guerini, Milano

Chapter 3
Digital Production

Abstract Contributions towards the end of the war were made by the prototypes of several innovative technologies, for example information technology with the first digital computers. The United States placed themselves at the forefront of technology transfer with structural development policies and strategies. Within 20 years they laid the foundations for the development of digital design and manufacturing in a broad range of industrial sectors, which would later be applied in the construction sector too. The peculiar needs of the aerospace industry for precision and flexible production made it a particularly receptive sector for innovative processes. In 1949, the US Air Force financed a proof-of-feasibility study of a three-axis milling machine operated by a computer. This technology took the name *Numerical Control* (NC). With NC, for the first time it became possible to describe individual elements of the design and manufacturing process in the formal language of science and technology, that is mathematics. The programming languages of NC have the declared aim of being intelligible to technicians and to computers which, by digitalising the design in bits and bytes, could therefore autonomously interpret it and translate it from the intangible domain into the physical object: the artefact produced "automatically" by the new machine tools. The designers themselves became masters of the production process, describing design in the new languages of information technology aimed at direct manufacturing without the intermediary of descriptive geometry any longer. This new class of technicians no longer just supervises, but also manages the procedures for the manufacturing process through programming languages. Numerically controlled machines automatically interpret and execute the instructions of NC language, to make the parts that, assembled together, result in a complete product. NC technology, initially adopted by large industries which were structured around the division and separation of skills and jobs, was adopted by small and medium-sized enterprises as soon as investment costs began to drop. SMEs directed the technology to the demands of their own "horizontal" organisations, with a high level of skills shared between a numerically small workforce.

L. Caneparo, *Digital Fabrication in Architecture, Engineering and Construction*,   51
DOI 10.1007/978-94-007-7137-6_3, © Springer Science+Business Media Dordrecht 2014

3.1 Origins

Probably for the first time in history, during the Second World War, aviation made a fundamental contribution to deciding the outcome of the conflict. The aerospace industries of the various combatant countries were asked to produce a great deal of equipment needed to replace important losses and if possible to extend the operating capacity of their respective air forces. During the conflict, air forces found they needed continual innovations in their strategies and aircraft.

> At no other time in history have science and technology known such rapid progress as in the years between 1939 and 1945. Such progress was most marked and significant in the field of the design of military aircraft, some of which were developed in response to needs barely imagined in the years before the Second World War (Jackson 2006, p. 6).

Based on experiences in the various theatres of war, the aircraft being produced were continually revised and changed and all new experiences were transferred to new versions and models. Zeitlin (1995) considers a crucial factor in the success of the US Air Force to be the speed with which it gathered new experiences on the field and quickly turned them into strategic indications for the conduct of operations and for the development of fighting materials.

The aerospace industry was required to produce large quantities of high quality, but also flexible, material:

> Established mass-production methods, such as those pioneered by the automobile industry, typically proved too rigid for the high level of uncertainty and rapid pace of innovation imposed by the war economy. Successful aircraft manufacturers therefore needed to find new ways of reconciling the high throughput of mass production with the adaptability of the craft workshop. The result was a proliferation of hybrid forms of productive organization, which anticipated in many respects more recent innovations in flexible manufacturing. (Zeitlin 1995, p. 49)

The rapid pace of innovation spurred the United States aerospace industry to experiment with and to employ new methods and technologies. At the end of the war, the American aerospace industry was the first to experiment with the integration of computers in industrial production. The first prototypes of electronic computers were the product of huge investments by the military in automating the deciphering of secret codes (Figs. 3.1 and 3.2).

England was at the forefront of the development of hardware and software: at Bletchley Park a group of exceptional theorists and technologists, the mathematicians Alan Turing and Max Newman and the engineer Tommy Flowers, with the collaboration of Sidney Broadhurst, William Chandler, Allen Coombs and Harry Fensom, made and began to use Colossus. This computer used thermoionic valves for its computations, much slower and more expensive than the transistor, which was only invented after the war and began to be commercially available in the 1960s. Colossus began operation in 1944.

In Germany, Konrad Zuse designed and made a computer for fighting purposes, whilst the United States began the construction of its own first digital computer only in 1943 (Goldstine 1972). This was the Electronic Numerical Integrator And

Fig. 3.1 Colossus, the first digital computer in history, operating at Bletchley Park in 1944

Fig. 3.2 Electronic Numerical Integrator And Computer (*ENIAC*) introduced in 1946. Goldstine estimated the cost of the project to be the equivalent of five million dollars in 1970

Computer (ENIAC). Despite the advantage gained by the English in the area of information technology, the industrial development took place in the USA. This may be ascribed to the integration and interchangeability between industries and the joined-up system of military financing and commissioning, a system that had been experimented with from the time of the foundation of the United States and was now fine-tuned.[1] Kenneth Flamm (1988) believes that the reason why the United States led the field was the unwillingness of European governments to fund research in the developing information technology industry and to acquire innovative technologies. In the post-war period, the success of industrial groups like UNIVAC and IBM made ever more powerful, reliable and economically competitive computers available on the market.

The application of electronic computers to industrial manufacturing is thought to have begun with John Parsons and Frank Stulen in Michigan around the end of 1940. The Parsons Corporation received a commission to make the wooden ribs in the rotor blades of one of the first helicopters produced on an industrial scale, the Sikorsky. When one of the blades failed, Parsons suggested making the stringers out of steel. The profile of the stringer was provided by Sikorsky by means of a sequence of 17 reference points. Parsons had to join the points to define the cutting template. He consulted Frank Stulen, the director of the Rotary Ring division at Propeller Lab. On being immediately hired, Stulen made a simulation of the structure of the blade using the computers of the time, which were programmed with punched cards. He obtained 200 control points for drawing the template. In place of the manual method, which needed every point to be plotted on a Cartesian graph, Parsons designed a completely automated system, in which the X and Y axes of a milling cutter were moved by electric motors controlled by a punched card reader (Olexa 2001). However, Parsons did not have the money to make the complete system.

The peculiar needs of the aerospace industry for precision and flexible production made it a particularly receptive sector for innovative processes. In 1949 the US Air Force financed a proof-of-feasibility study of a three-axis milling machine operated by a computer. This technology took the name *Numerical Control* (NC).

In the same years, General Electric developed and achieved the automation of machine tools by recording the manual operations of a machinist. The machine recorded on a tape each action carried out by the operator. Subsequently, by playing back the tape, the machine repeated automatically the sequence of actions, reproducing an unlimited series of pieces. In this way, a highly specialised worker, skilled in operating a traditional machine tool, can be replaced by a worker charged with supervising the production process automatically carried out by a machine.

Recording the sequence of operations makes it possible to acquire the whole specific craft skill of a process, comprising the choice of sequences of movements and operations, speed of processing, type and characteristic of the tool, positioning of the piece and so on.

It very soon became clear that the technology of Numerical Control had the potential for much more than just recording and carrying out sequences of operations, for the recording process continues to delegate the strategic choice of working processes to the

---

[1] See Sect. 2.1 "Uniformity and Standardisation" in this volume.

Fig. 3.3 Card-a-matic Milling Machine at the Servomechanism Laboratory of Massachusetts Institute of Technology

decisions of the specialist machinist who makes the recording. The influence of design on this procedure is only indirect, as it can neither control it nor improve it much (Kochan 1986; Callicott 2001; Noble 2011).

Following funding by the Air Force, Parsons Corporation invited the ServoMechanism Laboratory at the Massachusetts Institute of Technology to join the project in order to develop a prototype of a machine tool which would generalise the principles of numerical control. Within a year they made the "Card-a-matic Milling Machine" for three-axis contour milling controlled by a computer and programmed by means of punched cards (Pease 1952).

Tests on the Card-a-matic Milling Machine were very positive in terms of precision and repeatability compared with traditional machine tools. Several test pieces were processed, including a launching socket made out of a stock of aluminium alloy. The automatic processing required only three and a half hours, while the programming process and preparing the punched cards took 120 h, plus a further 80 h to fix the workpiece to the worktop (Fig. 3.3).

The Card-a-matic Milling Machine provides an example of the way numerically controlled machines work which is still valid today:

- the machine is a traditional milling machine, produced by Cincinnati Milling Machine Company, with three axes of movement; each axis is automated with its own servomotor;

- the individual motors are controlled by a digital processor, known as "The Director", which would later be called the control unit;
- the computer reads the program, in this case encoded on punched cards, interprets it as instructions and transmits it to the control unit.

The potential advantages for the production cycle were seen immediately: the reduction in down time and increase in productivity; uniformity in processing with greater precision standardisation and thus repeatability; the reduction in specialised workforce; the optimisation of waste materials; the increase in versatility through the capacity to use a single machine tool, a work centre, instead of several specialised machines such as boring, milling or drilling machines. However, the programming and the fixing of the workpiece were laborious and complicated.

The industry was unwilling to make the huge investments necessary for the industrial development of the technology for numerical control, so as to produce a radically new range of machine tools. In this circumstance the problem was again solved by the intervention of the government: in 1956, the US Air Force commissioned several firms to develop numerically controlled machine tools.

3.2 Automation and Technological Convergence

The potential advantages of numerically controlled machines for the aerospace sector were immediately obvious, but in order for the industry to be able to take advantage of them, they had to reach a "critical mass" through widespread diffusion of the technology. The problem was the high cost of the NC system because it brought together several technologies which at the time were cutting-edge, just out of the research laboratories. Servomotors, electronics and computers were then just beginning to address the problems of mass production, with the development of a new research domain, microelectronics.

The costs were prohibitive even for multinational aerospace companies. During the 1960s,

> the Air Force undertook to pay for the purchase, installation, and maintenance of over one hundred such machines to be installed at government expense in the factories of prime- and sub-contractors; the contractors, aircraft manufacturers, and their suppliers would also be paid to learn to use and maintain the new technology (Noble 1979, p. 329).

The total value of the investment sustained by the Air Force was more than 62 million dollars at the time.[2]

The Air Force insisted that all the machine tools adopt the standard programming system, based on the Automatically Programmed Tools (APT) language:

> The APT system was flexible, fundamental, a basic system for up to five-axis control. The Air Force loved it because of its flexibility; it seemed to allow for quick mobilization, rapid design change, and interchangeability between machines in a plant,

---

[2] Equivalent to an investment of more than half a billion dollars today.

between users and vendors, and between contractors and subcontractors. With these ends in mind, the Air Force pushed for standardisation of the APT system; [...] Before too long, the APT computer language had indeed become the industry standard, despite initial resistance within the plants of aircraft companies. Many of these companies had developed their own languages to program their N.C. equipment, and these in-house languages, while less flexible than APT, were nevertheless proven, relatively simple to use, and perfectly suited to the particular needs of the company. APT was something else entirely. For all its advantages, indeed because of them, the APT system had decided disadvantages.

The more fundamental a system is, the more cumbersome it is, and the more complex, the more skilled the programmer must be and the bigger the computer must be to handle the larger amount of information. In addition, the greater the amount of information, the greater the chance for error (Noble 1979, p. 331).

The consequence was an initially high investment, sustained by the government, which provided the incentive for industrial development in very innovative sectors like microelectronics, information technology and machine tools. Moreover it achieved the adoption of technologies within strategic industrial sectors, beginning with aerospace, soon followed by industry as a whole. A further effect was the rapid computerisation of US industry.

3.3   Computerisation of Industry

American industry had begun to acquire electronic computers principally for administrative purposes, for the management of accounts. The initiative of the US Air Force during the 1960s triggered a radical change in the organisation and management of the industrial process (Fig. 3.4).

Numerical control was based upon an entirely different philosophy of manufacturing. Here the specifications for a part, the information contained in the engineering blueprint, is broken down, first, into a mathematical representation of the part, then into a mathematical description of the desired path of the cutting tool, and ultimately, into hundreds or thousands of discrete instructions, translated for economy into a numerical code, which is read by the machine controls. The N.C. tape, in short, is a means of formally synthesizing the skill of a machinist [...] This new approach to machining was heralded by the National Commission on Technology, Automation and Economic Progress, as "probably the most significant development in manufacturing since the introduction of the moving assembly line" (Noble 1979, p. 327).

Controlling a machine tool by a computer is a relatively accessible process, first developed by Parsons and Stulen, then accomplished by the MIT Laboratory in less than a year. During the nineteenth century some levels of automation had already been developed and applied to machine tools: for example automatic loaders, stops, fixing and cams. These arrangements were aimed at simplifying and speeding up the tasks of the machinist.

Numerical control, particularly programming the operations of a machine tool, was something new, however. For the first time a single description, the NC code, synthesised the representations necessary for design and production.

Fig. 3.4 Example of the
APT II program. Symbolic
names are assigned to the
geometric primitives: JIM is
a point with coordinates 6
and 7, JANE is the
circumference of a circle of
radius 2 and centre JIM,
WALDO is the outline
obtained by a tangent to the
circles JANE and JUNE

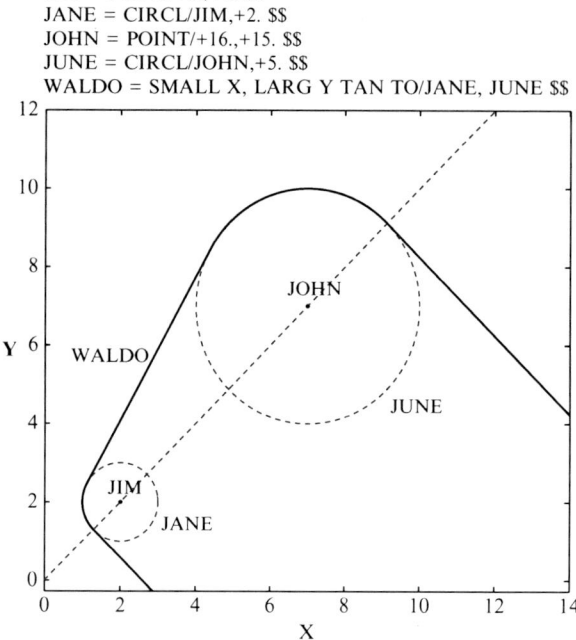

```
JIM  = POINT/+6.,+7. $$
JANE = CIRCL/JIM,+2. $$
JOHN = POINT/+16.,+15. $$
JUNE = CIRCL/JOHN,+5. $$
WALDO = SMALL X, LARG Y TAN TO/JANE, JUNE $$
```

The shape of the piece is formalised according to analytical geometry, broken
down into "planes, spheres, cones, cylinders, and quadric surfaces" (Ross 1978, p. 289),
defined by their dimensions and spatial coordinates.

> The objective is to instruct the machine tool to perform some specific operation, so that
> sentences similar to the imperative sentence form of English are required. [...] Declarative
> statements are also necessary. Examples of declarative sentences used to program a numerically
> controlled machine tool might then be of the form:
> "Sphere No. 1 has center at (1, 2, 3) and radius 4"
> "Airfoil No. 5 is given by equation [...]"
> "Surface No. 16 is a third order fairing of surface 4 into surface 7 with boundaries [...]"
> An imperative sentence might have the form:
> "Cut the region of Sphere No. 1 bounded by planes 1, 2, and 3 by a clockwise spiral cut
> to a tolerance of 0.005 inch"
> These sample sentences, although written here in full English text for clarity, are
> indeed representative of the type of language toward which the present study is aimed
> (Ross 1978, p. 293).

The APT programming language, which stands for "Automatically Programmed
Tools", established the syntax and the semantics which the programmer must use in
writing the code. Semantics associates symbolic names with primitives and operators, to make reading and writing easier for the programmer. They are names or
abbreviations in English: for example, Sphere, Profile, Surface. Syntax established
the rules for putting together the symbols defined in semantics. For example, the
primitive Sphere is followed by its coordinates, which must be separated with the
appropriate separator character.

3.3.1   From Universal Language to Numerical Control

> The idea of a 'language', the rules of operation of which apply to some 'atomic' elements defined in its realm, attracted also the attention of Monge. By the end of the eighteenth century he used these ideas in his formulation of descriptive geometry, a new branch of geometry dealing with the representation of three dimensional objects on a two dimensional support; for example, on a sheet of paper. From there, their shape could be easily 'read' by anyone, anywhere, knowing just the set of representation-rules formulated by Monge. We could say, by understanding Monge's 'language' (Ortiz 2010).

Gaspard Monge was a leading figure in the years of the French Revolution, during which he held several important positions. In 1792 he was made Minister of the Navy, but the following year he asked to be relieved of the assignment after having measured the impossibility of introducing effective changes in its erratic and inefficient organisation. During the invasion of France by Austria and Prussia, in order to satisfy the pressing need for warfare material, he contributed to improving the efficiency of the war industry through the diffusion of technological innovation.

Once the external dangers to the Republic were overcome, Monge's work turned to forming a new technical class capable of producing innovations and progress in French society. The Commission for Public Works, of which Monge was a member, gave birth to the École Polytechnique. Just 1 year after its foundation, the school was ready to operate. Monge played multiple roles: organiser, administrator and teacher.

> The descriptive geometry developed by Gaspard Monge (1746–1818) made available a graphic language and a conceptual tool for the treatment and design of spaces. The tool available in the field of analysis, the calculus of variations, seemed perfectly suitable for translating into mathematical terms the prescriptive problems typical of engineering practice (Lucertini et al. 2004, p. 29).

An important role of the technical class, formed directly by Monge through the École Polytechnique, was to improve "the working tool" (Valmalette 1957), adapting it to the evolution of production and construction processes.[3]

At the time of the foundation of the École Polytechnique, almost 50 % of the courses were dedicated to descriptive geometry (Gabetti 1968) according to explicit theoretical and practical aims with deliberately technological applicability.

> In *Géométrie descriptive*, practical and theoretical considerations were systematized and subsumed to a clear technological intentionality. To attain truth, mathematical precision was necessary in all disciplines. Monge also believed that a popularization of scientific methods and outlook was imperative for the advancement of industry. This would finally dispel the mystery concerning many manufacturing processes. He perceived his own work as a basis of the new *ars fabricandi* of technology, the "theory" of the new breed of engineers, whose only purpose was to make production more efficient (Pérez-Gómez 1985, p. 280).

---

[3] Perhaps the decisive element of the new organization for the study of construction is represented by the creation of the École Polytechnique: engineering so clearly superseded architecture, and at the same time, it established a closer relationship between the physical–mathematical sciences and technical applications, democratising the theoretical debate, and offered increasing scientific dignity to design, especially in relation to the great public works (Benvenuto 1981, p. 417).

The "universal language", and more generally the mathematisation of knowledge, are the means to change the relationship between production and technology, conceived in a new way from the second half of the 1700s. The Illuminists became aware that the great achievement of the *Moderns* was the capability to organise empirical knowledge according to rigorous methods. Diderot and d'Alembert's *Encyclopédie*, and Réaumur's *Description des Arts et Métiers* are the most notable examples of scientists' desire to interpret and classify traditional areas of knowledge.

The Encyclopedists believed that scientific methodology can provide the way to interpret arts and craft know-how and to formalise it into a new structure of knowledge. Denis Diderot in his "Preliminary Discourse" explains stage by stage the methodology to be used in describing each art and craft.

> The first stage consists of laying out a very detailed description of existing techniques. [...] There are five points to analyse: firstly the materials used and where they come from, their preparation, their good and bad qualities; secondly the main working processes; then the tools and the machines used in them; then the type of workforce and the main operations involved in production, and finally the terminology. The heading 'Arts' provides some extra background indications about the study of machines. Both with complicated ones and with ones which carry out the most basic procedures, it is necessary to begin by analysing their effect before considering the machine itself, or else the description of the machine must come immediately before that of the effect (Picon 1992).

The plates in the *Encyclopédie* aid the analysis and description of the tools and machines and their specific use in each craft. Description of the different crafts is more problematic. Diderot experienced various difficulties in the practical outworking of his "five points":

> I remember one craftsman, to whom I believed I had explained exactly what I needed to describe his craft of carpet making, something which would have required a short text and a table of drawings; he brought me from ten to twelve tables full of diagrams and three thick folio-sized notebooks, enough to fill one or two volumes in duodecimo.
>
> Another, in contrast, to whom I had explained exactly the same rules and whose production activity was one of the most complex in terms of products made and materials used, brought me a short list of words without definitions, without explanations, and without diagrams, firmly assuring me that his craft had nothing else to it.

The *Encyclopédie* addressed and solved the variety and peculiarities of the single areas of knowledge through breaking down the process into acts and operations;

> showing the acts and the operations, before reassembling the production processes, the creators of the Encyclopedia move forward to a true reproduction of the arts and crafts [...] At this very moment we are watching the birth of a new rationale, an analytic rationale in the function of the revindication of the use of analysis in the technological area, whether it be in mathematics or aimed at human activities (Picon 1992).

Descriptive geometry has been adopted as the language through which engineers and architects communicate with mechanics and builders, a language that each worker with any responsible role in the process needs to know. It is a language shared by all, as opposed to specialised dialects and jargons, peculiar to each craftsmanship and profession – a graphic "universal language" (Belofsky 1991) that has replaced oral traditions, which today have become obsolete, almost lost and forgotten (Dupire et al. 1985, p. 66).

3.3.2 *Language of Numerical Control*

In September 1956, during the kickoff meeting of the project that developed the APT language, the priorities were declared:

> First of all we want to establish that the problem is one of language. Then, what kind of language? Well, we want our language to be just between the man and his problem, independent of the particular machine tool that is going to be used but [...] we do have to compromise somewhat on this ideal language. In other words, our language will be influenced by the particular tool and by the particular computers to some extent, but our goal is to have it as nearly independent of these quantities as possible (Ross 1978, pp. 289–290).

APT is a language for describing the specifications of operations and of machine tools, the trajectories of the tools relative to the geometry of the workpiece according to parameters dependent on the feed and rotation speed, the tool, the material and the wear of the cutter.

These are unwritten rules, acquired with the experience and skill of the craftworker and the machinist.

The compilers of the Encyclopaedia undertook the breaking down of working processes in the arts and crafts with detailed descriptions of movements and operations along with narratives as well as tables and drawings. The analysis and description of working processes became a line of research into industrialisation. The methods of Taylor and his circle and successors may vary in detail among themselves, but they all have in common the desire to put time and motion studies at the centre of interest.

The purpose of research in scientific management is

> Analyzing the motions of the workmen in the machine shop [...] all the operations for example which were performed while putting work into or taking work out from the machine (Giedion 1948).

With APT, it became possible for the first time to describe the individual elements of the process in the formal language of science and technology: that of mathematics.

The new programming language had the declared aim of being intelligible to technicians and to computers which, by digitalising it in bits and bytes, could therefore autonomously interpret it and translate it from the intangible domain of design into the physical object; the artefact produced "automatically" by the new machine tools.

The very etymology of the neologism "Numerical Control" explains the significance of the innovation of machines controlled by the analytical language of mathematics, written by a new class of technicians (programmers) in accordance with the emerging language of information technology. It was necessary that this language be intelligible to the people it was aimed at, as was clearly stated at the initial meeting of the APT project.

The people it was aimed at were engineers, architects and designers in general, whose functions and skills were rapidly changing. Before the project, the relationships between design and production had been carried out through drafting, the universal language.

Fig. 3.5 This illustration from 1910, entitled *Vision of year 2000*, shows how progress was imagined: the architect from his operator panel has full and direct control of the construction process

> He [Monge] perceived his own work as a basis for the new *ars fabricandi* of technology, the "theory" of the new breed of engineers, whose only purpose was to make production more efficient (Pérez-Gómez 1985, p. 280).

For example:

> The company [Ford Motor] produced and maintained drawings of every part of the Model T, every special tool, jig, fixture and gauge used in its production, and every master gauge used to check these special devices. Drawings served as the ultimate authority in Ford production, for they specified dimensions, tolerances, gauging points, materials (including shear strength and other metallurgical specifications), and finishes. Used by the design, tool, engineering, and inspection… (Hounshell 1987, p. 272).

Later the designers themselves became masters of the production process, describing design in the new languages of information technology aimed at direct manufacturing without the intermediary of the "universal language" any longer. This new class of technicians no longer just supervises, but also manages the procedures for the manufacturing process via the programming languages. Numerically controlled machines automatically interpret and execute the instructions of APT language in order to make the parts that upon assembly result in a complete product (Fig. 3.5).

The completion of the transition to total control of the production process happened within the span of a couple of decades. With the rapid spread of systems for numerical control in industry, from the start of the 1970s, the experience of APT was to converge in the international standard RS274D, developed by the Electronic Industries Association and also known as G-Code, now very widespread in the USA, whilst the standards adopted in Europe were ISO 6983 and DIN 66025. In industrial management, programming languages required a new technical staff with the IT skills indispensable for writing computer code, and the skills in mechanics and manufacturing necessary for collaborating in the design process.

The 1970s saw a parallel extension in research and development in Computer-Aided Design (CAD) systems, which provided a graphic interface for creating geometric representations. They offered a virtual design table that was much more intuitive and suitable for the experience of designers and draftsmen than the syntax and semantics necessary for programming languages.

The evolution of NC languages was towards graphic interfaces. It was the organisation of the design process that made it possible to add to the geometric description of the piece, already done in CAD, the definition of the characteristics useful to the work cycle; the characteristics of materials and NC machine tools. The new software was called Computer-Aided Manufacturing (CAM) and either works as part of the CAD system or includes its own functions for the geometric modelling of a workpiece.

From the end of the 1960s to the end of the 1970s, numerical control expanded to an ever growing number of manufacturing processes, keeping pace with the spread to different industrial sectors, until it carried out all stages of the production cycle, including managing the raw materials, processing and assembling components. Automated systems for the entire production process are called "flexible manufacturing" and make use of increasingly versatile new technologies, including machining centres, robots and computer networks.

The large industrial groups prioritised flexible manufacturing systems in their research and development into innovative methodologies and working models. For example, Toyota aimed the characteristics of flexible manufacturing at the Japanese structure of industrial relations and introduced "just in time" production. By reducing production time, Toyota succeeded in making pieces to order by the client rather than making predetermined lots of parts to be stored in the warehouse waiting for orders. Automation supports the complete process from the order to shipping so that design and production are two stages in the overall integration.

Further developments at Toyota created "lean manufacturing"; the design of the components of a vehicle is simplified so as to optimise flexible production in real time as regards assembly and warehousing. Toyota is still experimenting with methods of total quality management, replacing former methods of quality checking which, at the completion of the production process, determine whether the component is up to standard or must be discarded. Innovative methods of quality control are carried out during the course of the process. Any quality problem is tackled at the level of the individual process when it surfaces (Womack et al. 1990).

3.4 Industrial Districts

The large scale production of some fundamental subsystems of numerical control (workstations, servomotors), considerably reduced the initial costs of investment in NC technology. In the meantime, increasingly flexible machine tools were developed, first of all machining centres and robots, which could replace several earlier single-purpose machines (Bralla 2007; Groover 1996; Overby 2011). The combined action of these two factors gave even small and medium-sized enterprises (SMEs) access to

Fig. 3.6 Robotised assembling of bricks for the walls of Gantenbein Vineyard. At Architecture and Digital Fabrication, ETH Zurich, the elements of the façade were fabricated from standard bricks. The industrial robot has applied the adhesive and positioned each brick

the technology of numerical control. From the 1970s, SMEs began to invest in NC, partly to replace obsolete machinery, partly as innovation. Numerical control, perhaps even on one machine, was introduced in the context of high or even very high manufacturing skills. Experience and skills are indispensable for guaranteeing productive flexibility in medium-sized and small organisations, in which a small number of employees must carry out the whole process, assuring high quality standards, often higher than those of the larger industries. In SMEs, the functional separation between design and the production process is minimal, if it exists at all. Even in the case of separation and specialisations of functions among different SMEs, their size facilitates direct interpersonal relationships. When two or three SMEs collaborate in the design and production of a product, there is usually a continual flow of know-how between nominally different processes and organisations, which in practice however are pragmatically interconnected in managing a common project.

In this context, it has turned out to be logical to transfer the knowledge capital that each SME has consolidated into the numerical control technologies, with innovative outcomes (Fig. 3.6).

3.4.1 Horizontal Integration

Compared with the organisation of the processes in a large industry, the very flexibility of the SMEs comes from a small number of skilled workers. SMEs can promptly adapt production to the changing demands of the market. The use of numerical control technology accentuates flexibility of production, in that NC machines can carry out several functions and can be adapted to different processes since they can be rapidly reprogrammed. In the context of SMEs, NC machines guarantee productivity even for small volumes of production.

At the lower limit of sizes in the construction industry, we find companies with fewer than nine workers, called micro-enterprises. For example, woodworkers and ironworkers tailor the details of their production of window frames and fixtures for each specific project. Many of these enterprises have invested in machining centres, whose flexibility makes it viable to produce even small series of a few dozen pieces. On the other hand, many of these enterprises work substantially by 'catalogue' or rather on a limited number of types of product, whose NC programs were not infrequently set up when the machine was bought and do not change over time. The reason is that few of these micro-enterprises have within their staff the NC skills required to obtain the full range of achievements possible, getting the best out of machining centres or other numerical control technologies.

Alongside these are enterprises, in some cases also small or very small, which because of a technological culture are capable of customising their production to the specific requests of the client, often the construction company or the designer. In the interaction, almost always direct and interpersonal,[4] an exchange of skills between design and production takes place. The result is the capacity to produce batches ranging from small to large according to the client's design and requirements. The highly skilled workers of these enterprises use NC technology creatively, and cross-fertilise it, case by case, within sometimes traditional and even craft processes.

The success of these SMEs shows the principle of technology transfer, under which a technology born and used in one industrial sector can be acquired and adapted by a different one.

> A further type of innovation regards the transfer of innovative elements (objects, materials, equipment, services etc.) from one sector to another; in this case the innovation already exists and migrates, suitably adapted and changed, to another working context (Sinopoli and Tatano 2002).

NC technology, initially adopted by large industries, at the time structured around the division and separation of skills and jobs, was received by small and medium-sized enterprises as soon as the investment costs became to drop. SMEs directed the technology to the demands of their own "horizontal" organisations, with a high level of skills shared between a numerically small workforce.

3.4.2 Vertical Integration

The presence in a particular area of enterprises which succeed in profitably sharing their skills so as to impact on product innovations in some cases starts off a process of accumulation and aggregation. In a virtuous circle, the activity of nearby

---

[4] The architect [Carlo Scarpa] established relationships for these works with trustworthy craftsmen: the metalworker, the woodworker, the plasterer, [...] He used to discuss it with the craftsmen and would go back again and again to check and comment on the state of the work and the results (Crippa 1984).

enterprises specialising in a production chain attracts, or makes it easier to meet, other enterprises willing to collaborate in the process. For example, in the sectors of stone or wood-working, when the enterprises keenest on innovation began to introduce numerical control technology, specific skills in programming NC or CAM were virtually absent in the geographical area. This spurred neighbouring small enterprises, which perhaps were already working in the field of information technology, to acquire the necessary specialist skills.

The result was mutual collaborations which, in some cases, influenced real innovations in the sector in terms of products or services. This makes an area more specialised and more attractive for investment, starting off a process which contributes to the creation of an industrial district (Bagnasco 1977).

Brian Arthur (1990) and Paul Krugman (1991) define an industrial district as the capacity to circulate positive returns or network externalities. In practice, in a cluster knowledge begets knowledge. Carabelli et al. (2008) believe that the expertise acquired by a district is a multiple of the sum of the individual parts and that the sharing of skills between the enterprises is not proportional to the number of workers or to market share.

> The significance of the spatial or geographical dimension of innovation has been demonstrated by research aimed at understanding the successful development of innovation in specific locations such as the M4 corridor in the UK and Silicon Valley in California. This research has highlighted the importance of concepts such as industrial districts, innovative milieux and regional innovation networks [...]. The emergence of regional concentrations of innovation has reinforced the view that innovation is a collective process, dependent on many different interactions between an organization and its external environment, which includes suppliers, customers, technical institutes, training bodies, technology transfer agencies, trade associations and other government agencies.
>
> In line with the neo-Schumpeterien research, the current debate also supports the idea of adopting a multidisciplinary and integrated approach in trying to evaluate effectively the innovative potential of an organization, an industry, a region or a nation. The value for such an integrated approach is that it implies that innovation should be seen not as a separate activity, but as a whole, integrating the organization with its entire external environment (Jones and Saad 2003).

Architecture, engineering and construction (AEC) has widely adopted project-based forms of organisation, i.e.

> working together through agreeing mutual objectives, devising a way for resolving any disputes and committing themselves to continuous improvement, measuring progress and sharing the gains (Egan 1998).

The mechanisms that might contribute to explaining the diffusion of innovation in project-based industries have been studied for instance in AEC (Arditi and Tangkar 1997; Eccles 1981; Henderson 1996, pp. 359–375; Levitt et al. 2012, 2013; Taylor and Levitt 2007; Tatum 1989), biotechnology (Al-Laham and Amburgey 2011, pp. 323–356; Barley et al. 1992; Brown and Hendry 2006; Phene et al. 2006; Zucker et al. 1996), healthcare (Cooke 2001; Esteve et al. 2012; Luke et al. 1989), movie (Lampel 2011, pp. 445–466), pharmaceutical (Kennedy 2008; Orsenigo et al. 2001; Zeller 2002), and software (Almeida and Kogiit 1999; Egorova et al. 2009, pp. 110–111; Isaksen 2004) sectors. These studies have argued that project-based industries consolidate a network of knowledge-intensive collaborations based on establishing and renewing relationships frequently, meeting the changeable requirements of clients, markets, procurements, tenders and so on.

Al-Laham and Amburgey (2011) point out that a firm's potential to innovate is proportional to its ability to create and to dissolve relationships as an effective capability to bring to the firm new project opportunities and access to inventive knowledge.

> Given the fundamental importance of knowledge for innovative success, we argue that biotech firms who access and exploit knowledge from their project-alliance or project-networks will increase their innovation speed significantly. More specifically, we state that firms who are connected to external knowledge sources will develop their learning capabilities faster than unconnected firms, which eventually leads to an increase in their rate of patenting.

In the area around Amsterdam and Rotterdam a district of architectural practices has become established, the so-called "Superdutch architecture" generation, encompassing among others Erick van Egeraat, Kees Christiaanse, Meccanoo, MVRDV, Neutelings & Riedijk, NOX, Rem Koolhaas, UN Studio, West 8 and Wiel Arets (Lootsma 2000). The proximity and the highly urbanised surroundings help drive innovation, efficiency, and flexibility:

> The case of strong-idea architects in Amsterdam and Rotterdam displays the spatial concentration; a specialized labour market and we also see the contours of a 'dedicated local institutional infrastructure' with both formal (e.g. Berlage Institute, NAi, grant system) and more informal components (e.g. Koolhaas' OMA as an informal postgraduate school). Spatial proximity may generate important emergent effects such as mutual learning and cultural synergies through knowledge spillovers, which underpin sustained innovation (Kloosterman 2008).

Despite the fact that in this area it seems that no direct inter-firm collaboration or linkages exist, the district acts as a network through indirect sharing of knowledge. One primary factor is access to a highly skilled and specialised labour pool, which constitutes a crucial factor for every innovative industry. The high labour mobility offered by the spatial concentration results in a spill-over of knowledge. Further means are offered by the cultural and institutional context built around dedicated infrastructures, e.g. several universities including The Netherlands Architecture Institute, The Netherlands Architecture Fund, the Berlage Institute.

> The formal educational institutions are very important bridges or conduits of learning in the architectural field. Nearly all of the interviewed strong-idea architects were at the moment of the interview working as a teacher or lecturer at a school of architecture or had worked as such. Combining their work as architects with (part-time) teaching jobs helps young architects to survive in the first difficult phase as head of an independent practice. It enables them and other more established to put their views forward, to find new, talented and suitable workers for their own firm, meet other lecturers and also to keep in touch with new views and fads in architecture (Kloosterman 2008).

The Netherlands district of architectural practices has a horizontal integration mainly because it comprises firms with similar specialisation. With respect to vertical integration,

> the Dutch construction cluster is a fairly traditional cluster, a project-based cluster in which informal networks and relations are important and knowledge management practices need further development. Proven technology and organizational concepts prevail as well as risk-averse clients, detailed procurement and a clear focus on prices and costs.
>
> Current challenges include scarcity of space, requiring creative solutions; more demanding end users; a changing role for government; and increased competition on construction markets. The latter comes particularly from non-construction firms in the cluster that have started to integrate further backwards and forwards in the value chain. It is against this rather complex organizational background that innovation in the construction cluster has to be examined. The functioning and innovation style, or more likely innovation styles, in the various sub-clusters of the construction mega-cluster, are specific to the cluster. In well-established, mature clusters in particular, "recipes" for innovation cannot be copied or changed overnight.
>
> As in most other clusters, innovation in the construction cluster is a multifaceted phenomenon involving clear technological, organizational and market aspects. The quantitative data presented in this contribution mainly focus on technological innovation in the Dutch construction (mega-) cluster. Innovation in the overall construction cluster was found to be higher than for the pure construction industries at the core of the cluster. The latter score considerably lower in terms of (mainly technological) innovation, innovation networking and the degree to which innovation is thought to have contributed to added success on sales markets. The same data show that the manufacturing and service firms active in construction markets perform much better. For their (technological) innovation, construction firms are dependent on innovation in manufacturing firms both upstream and downstream in the value chain. This need not be a problem if the ability to absorb out-of-industry technology and the links with end users are well developed. It is precisely these factors that qualitative studies highlight as being important barriers for innovation. Such barriers include the need to co-ordinate disciplines and the various phases in the building process, a limited sensitivity to the needs of end users, detailed procurement and the current lack of market pressure to move forward with innovative products. However, these qualitative assertions, which mainly point to a considerable need for organizational and market innovations, could not be substantiated quantitatively using standard innovation data as such data mainly measure technological innovation (Den Hertog and Brouwer 2001, p. 203).

Firms and companies in the construction industry rely increasingly on the division of tasks via contracting and sub-contracting and on the shift in the balance of power between clients, architects and contractors. On the other side, clients especially in larger projects have gained greater contractual power as well as main contractor industries. These mayor players have strengthened their relationships into network-ties that tend to shift away from project-based alliances.

Several authors consider project-based networks an effective means of establishing effective collaboration with other organisations in order to deal with continuous technological innovation. Adaptability is crucial in evolving sectors and requires the rapid translation of knowledge from a technological invention or a groundbreaking design into a marketable innovation.[5] Levitt (Taylor and Levitt 2005, pp. 247–256) distinguishes between systemic and incremental innovation:

---

[5] Such as in the considered AEC, biotechnology, healthcare, movie, pharmaceutical, and software sectors.

Incremental innovations are those that reinforce the existing product or process and provide a measurable impact on productivity (e.g., transitioning from "stick-built construction" to the use of prefabricated "wall trusses" in homebuilding). In the case of incremental innovations, productivity for individual components can increase while overall productivity may increase, decline, or remain unchanged. Systemic innovations, on the other hand, refer to innovations that reinforce the existing product but necessitate a change in the process that requires multiple firms to change their practice. Systemic innovations typically enable significant increases in overall productivity over the long term.

The districts may be considered as organisations which collectively possess skills and which, in some cases, succeed in bringing systemic innovations to fruition. The economists Abernathy, Christensen, Clark and Fujimoto (Afuah 2009) have considered "disruptive" technological innovations:

> A disruptive technology is a superior alternative to the currently dominant know how, whose potential escapes the most masterful producers and users of the dominant method precisely because their experience teaches how to improve on what they already know (Sabel 2004).

For this reason, disruptive technologies are developed, first emerging from secondary markets and then consolidated worldwide. Significant examples are the arc welding machine; diamond thread for cutting; hydraulic earth movers or, as stated, the *lean production* method.

The National Centre for Italian Districts monitors 101 districts. Classification by sector of specialisation indicates the prevalence of three sectors, respectively clothing with 31 districts, automation and engineering with 26, building and furniture with 22, to which may be added paper and publishing with two and culture with one (Osservatorio Nazionale Distretti Italiani 2011). Although the Italian districts have been and still are closely studied, if only for the interest in understanding the mechanisms useful for exporting the methodology in other areas, it is difficult to recognise a common origin.

In general, the initial stages of districts are characterised by productive specialisation, for example stone at Carrara, chairs at Udine, furniture at Livenza and Murge, tiles at Sassuolo, ceramic sanitary-ware at Civita Castellana and plywood in the Triveneto and in Trentino. Around these districts have grown up enterprises specialised in machine tools and in information technology. In some cases, consultancies and the production of machinery have supplanted manufacturing because of global competition, as a result of which it can be more convenient to outsource production in emerging countries due to lower costs of the workforce or of the raw materials. Examples include the parquet and stone cladding industries.

With raw materials becoming ever more scarce and expensive, a major re-engineering of the products was achieved, with less use of expensive materials, in exchange for greater added value in processing, for example through prefinished integrated systems of joints. In these circumstances, the clusters are becoming exporters of know-how and technology.

The district of Carrara is characterised by a historical production specialisation in the marble sector which includes the complete supply chain, from quarrying to processing and working the natural stone, alongside which manufacturers of

Fig. 3.7 The number of industrial districts by region monitored by the Italian National Centre

machine tools and specialist consultancy firms have grown up. International competition from countries quarrying stone, who themselves became producers of finished products, brought into question the role of the Apuan district as a manufacturer of semi-finished and finished products and exporter of technologies (Fig. 3.7).

The district responded to the changed economic context by implementing policies aimed at developing the production and commercial supply chain as distinct from the mere quarrying capacity. The intention is to promote the capacity for industrial processing, including innovative production lines, for example ultra-thin layers of marble and composites, through technological innovation. The current trend of the Apuan cluster shows a picture of light and shade as it emerges from the sector survey published in 2010 by the Internazionale Marmi e Macchine of Carrara.

> The figures certainly show a comforting situation for the sector in the indication of an upturn and the consequent increase in stone exports, but unfortunately absolutely nothing to get excited about in the composition of the entries on the budget items […] in the course of 1 year Italian exports of blocks of marble went from 1,075,010 to 1,321,127 tonnes with an increase in value of 27.15 %, that of worked marble held steady at around 870,000 tonnes, with an increase in value of 8.4 %. […] the increase in sales of blocks is matched by the reduction in sales of worked stone. We can discover from the destinations that we have sold most of these blocks to the Chinese, from which we have banked 83 of the 104 million euros of exports […] these same Chinese have transformed them with cheap processes into floors and cladding then sold them to customers which once bought exclusively from Italian companies (Marabelli 2011).

The export of machines and equipment for stone working in 2010 reached 851 million Euros, with an increase of 31 % over the previous year, in which a dip of 32 % was reported.

The sector studies confirm the general modest propensity of the districts to invest directly abroad, because they tend to remain rooted in the territory where they grew up. However, the economic trend is determining a selection between the enterprises which in the districts succeed in establishing strong relationships with the various international markets. Significant examples are in the sectors of ceramics and of lighting components (Caroli 2007; Osservatorio Nazionale Distretti Italiani 2011). The capacity of the districts resides in the forms of collaboration between different companies. The nature of the economy is continually evolving. The districts are required to innovate in order to improve the capacities for jointly designing complex products, striving towards co-operation on a scale beyond the purely local. Particular attention must be paid to the development not so much of the product itself, something at which the districts have already proved excellent, as of designing the complete process which generates value from the product.

> The names of these medium-sized enterprises are well-known. In the field of the food sector Lavazza, Granarolo and Antinori; in the field of fashion Diesel, Max Mara, Geox, Calzedonia-Intimissimi, Marcolin, Ferragamo, Tod's; in the furniture sector companies like Natuzzi, Fantoni, Frau and Marazzi; in the automation-engineering sector the leaders of the growth of Italian products are for example Danieli, Sacmi, Merloni, Permasteelisa, and De Longhi. In many cases firms are linked to experiences of successful districts; in others, these medium-sized enterprises have been able to go their own way in this area. They all, however, have in common the capacity to field competitive autonomous and original strategies, sometimes very different from the management models which have characterised the traditional enterprises in the districts. They have not followed the crowd; they have made their own ways.
>
> Often with great success. […]
>
> Globalisation represented one of the distinctive aspects of the strategies put in place by the new leading enterprises. The projection outside the local system was achieved along several different development lines: upstream, through the delocalisation of some manufacturing activities and the recourse to new suppliers for raw materials and services; downstream, through acquisition and consolidation in foreign markets. The globalisation of production – it is right to underline this – constituted one of the most obvious breaks with the traditional pattern of the districts. The leading Italian export companies knew how to become active links in the global value chain, finding their own place within the global processes of the reorganisation of manufacture.
>
> The path to international opening of the medium-sized enterprises was translated into economic value when it was joined to a process of internal reorganisation. The medium-sized enterprise which worked successfully on the international markets knew how to redefine business processes through the use of new technology, made investments in the field of applied research and design (often in collaboration with universities and research centres), and put in place new communication and brand-development policies. These changes contributed to giving a better quality of management to the medium-sized enterprises and reinforcing their competitive advantage: globalisation paid when it could rely on a structured organisation (Micelli 2011).

A crucial factor for the districts is the ability to innovate processes, the structure of relationships and industrial collaboration through new applied technologies.

This requirement depends on the ability of research centres to operate from an agenda of priorities and themes. An excellent example is the creation of Silicon Valley. Before the microelectronics industry began to emerge, the Provost of Stanford University, Frederick Terman, had traced a clear plan for development.

Under his direction, the university supported the creation of specialised small enterprises which put research development into practice at the industrial level. It thus became a model of strategic planning in research and development, according to which

> several enterprises took up their headquarters there or were created, interacted with the laboratories of Stanford, thanks to substantial federal funding, and put into practice incisive development strategies (Kenney 2000).

A further requirement is the ability to create structures which can translate technological innovation into the languages and operating procedures peculiar to the forms of collaboration of each cluster. An interesting example of the integration of digital technologies in design, production and construction is the spin-off company of the Polytechnic of Zurich, designtoproduction GmbH:

> designtoproduction implements digital process chains from design to production based on parametric CAD-models and offers consulting services for parametric planning, detailing, optimization, and digital manufacturing.
>
> Our interdisciplinary team integrates specialist knowledge from various fields to help architects, designers, engineers, and manufacturers bridge the gap between idea and realization. Our services bring non-standard architecture to life [...] designtoproduction collaborates with fabrication experts to manufacture complex geometries on computer controlled tools.
>
> CNC machines make it possible to fabricate individual components almost at the cost of mass production. But while the price for machining stays the same, the effort for planning and logistics rises with the number of different parts. And when it comes to the thousands, every second lost per part adds up to man-hours of work in the project.
>
> We set up digital production chains from the design right to the machine, automating the generation of plans and fabrication data and ensuring quality and flexibility at the same time.[6]

The history of this spin-off is closely related to developments within academic research and to developments in the design and construction of innovative buildings. In 2002 Fabian Scheurer became a researcher at the Zurich university of science and technology ETH. In 2005 he, with Christoph Schindler and Markus Braach, founded the CAAD research group designtoproduction within the Department of Architecture. In 2002, Arnold Walz took part in the construction of the Mercedes Benz Museum, designed by UNStudio, as a consultant in geometric modelling. In the following years, designtoproduction collaborated in numerous projects which applied the achievements in research. Collaboration in the design of complex buildings culminated in the creation of Freeform Timber, a joint venture between designtoproduction, the engineering firm SJB Kempter Fitze and the Blumer-Lehmann company, which had over a century of experience in wood working.

> The recent experiences of the districts demonstrates that the need to put into practice strongly innovative projects determines a push towards the sharing of projects between companies (Fontana 2011).

---

[6] http://www.designtoproduction.ch.

The experience of the Quadrilatero Marche Umbria S.p.A., a public company in design and construction, shows how organisations and resources for innovation in the sector of infrastructures (which has always been crucially important at both local and national levels) can be developed from within the local area. Quadrilatero S.p.A.

will allow an infrastructural and logistical gap to be overcome, to the advantage of the community and of the industrial districts (Romozzi 2010).

The company structure and that of the governance of the projects is inspired by the principle of public-private collaboration through the financial mechanism of value capture, which is generated by managing the infrastructures built. Quadrilatero S.p.A. links innovative principles of governance and funding with cooperation with local research centres. It commissioned the spin-off company of the Università Politecnica delle Marche, SmartSpace, to develop technology to support quality control of construction materials. SmartSpace has filed patents for a system of monitoring the environment and buildings with networks of wireless sensors. In SmartSpace, the tight interconnections between research and development is exhibited by the managerial structure, which has Prof. Mario De Grassi as President and Prof. Berardo Naticchia as business development director.

The development of ICT opens the door to the achievement of "Smart" intelligent products, able to interact with the context thus increasing the possibility of developing a physical connection between the real world and that of computers (promoting a structure of interconnections which in the jargon of ICT is called 'Internet of Things') (Naticchia and Cacciaguerra 2011).

SmartSpace for Quadrilatero S.p.A. has made a network system to track samples of concrete during the complete process, which extends from blending to casting and storing individual samples of material. A unique identifier is encased within each concrete sample. By tracking this identifier, it becomes possible to use ICT to trace periodic checks and quality controls.

References

Afuah A (2009) Strategic innovation: new game strategies for competitive advantage. Routledge, New York

Al-Laham A, Amburgey TL (2011) Staying local or reaching globally? Analyzing structural characteristics of project-based networks in German biotech. In: Cattani G et al (eds) Project-based organizing and strategic management, vol 28, Advances in strategic management. Emerald, Bingley

Almeida P, Kogiit B (1999) Localization of knowledge and the mobility of engineers in regional networks. Manag Sci 45(7):905–917

Arditi D, Tangkar M (1997) Innovation in construction equipment and its flow into the construction industry. J Constr Eng Manag 123(12):371–378

Arthur B (1990) Positive feedbacks in the economy. Sci Am 262:92–99

Bagnasco A (1977) Tre Italie: la problematica territoriale dello sviluppo italiano. Il Mulino, Bologna

Barley S, Freeman J, Hybels R (1992) Strategic alliances in commercial biotechnology. In: Eccles R, Norhia N (eds) Networks and organizations. Harvard Business School Press, Boston

Belofsky H (1991) Engineering drawing: a universal language in two dialects. Technol Cult 32(1):23–46

Benvenuto E (1981) La scienza delle costruzioni e il suo sviluppo storico. Sansoni, Firenze

Bralla JG (2007) Handbook of manufacturing processes: how products, components and materials are made. Industrial Press, New York

Brown JE, Hendry C (2006) Organizational networking in UK biotechnology clusters. Br J Manag 17(1):55–73

Callicott N (2001) Computer aided manufacture in architecture: changing the craft of design. Architectural Press, Oxford

Carabelli A, Hirsch G, Rabellotti R (2008) Italian SMEs and industrial districts on the move. Where are they going? In: Haar J, Meyer-Stamer J (eds) Small firms, global markets: competitive challenges in the new economy. Palgrave Macmillan, Houndmills, Basingstoke

Caroli MG (ed) (2007) Il processo di internazionalizzazione delle piccole imprese. FrancoAngeli, Milano

Cooke P (2001) Regional innovation systems, clusters, and the knowledge economy. ICC 10(4):945–974

Crippa MA (1984) Carlo Scarpa: Il pensiero, il disegno, i progetti. Jaca Book, Milano

Den Hertog P, Brouwer E (2001) Innovation in the Dutch construction cluster. In: OECD (ed) Innovative clusters: drivers of national innovation systems. OECD Publishing, Paris

Dupire A et al (1985) L'architettura e la complessità del costruire. Clup, Milano. French edition: Dupire A et al (1981) Deux Essais sur la Construction. Mardaga, Bruxelles

Eccles R (1981) The quasifirm in the construction industry. J Econ Behav Organ 2:335–357

Egan J (1998) Rethinking construction: report of the construction task force on the scope for improving quality and efficiency of UK construction. The Stationery Office, London

Egorova E, Torchiano M, Morisio M (2009) Evaluating the perceived effect of software engineering practices in the Italian industry. In: Qing W et al (eds) Trustworthy software development processes, lecture notes in computer science. Springer, Berlin/Heidelberg

Esteve M, Ysa T, Longo F (2012) The creation of innovation through public-private collaboration. Revista Española de Cardiología (English Edition) 65(9):835–842

Flamm K (1988) Creating the computer: government, industry and high technology. Brookings Institution, Washington, DC

Fontana D (2011) Il futuro che ci aspetta. Flessibilità, innovazione, partnership. In: Osservatorio Nazionale Distretti Italiani, Secondo rapporto dell'Osservatorio Nazionale Distretti Italiani. Federazione dei Distretti Italiani, Mestre

Gabetti R (1968) L'insegnamento dell'architettura nel sistema didattico franco-italiano 1782. Quaderni di studio, Torino

Giedion S (1948) Mechanization takes command, a contribution to anonymous history. Oxford University Press, New York

Goldstine HH (1972) The computer: from Pascal to von Neumann. Princeton University Press, Princeton

Groover MP (1996) Fundamentals of modern manufacturing: materials, processes, and systems. Prentice Hall, Englewood Cliffs/London

Henderson R (1996) Technological change and the management of architectural competence. In: Cohen MD, Sproull LS (eds) Organizational learning. Sage Publications, Thousand Oaks/London

Hounshell DA (1987) From the American system to mass production, 1800–1932: the development of manufacturing technology in the United States. Johns Hopkins University Press, Baltimore

Isaksen A (2004) Knowledge-based clusters and urban location: the clustering of software consultancy in Oslo. Urb Stud 41(5–6):1157–1174

Jackson R (2006) Aviazione della seconda guerra mondiale. Evoluzione, armi, caratteristiche. L'airone, Roma

Jones M, Saad M (2003) Managing innovation in construction. Thomas Telford, London

Kennedy A (2008) Pharmaceutical project management. Informa Healthcare, New York

Kenney M (ed) (2000) Understanding silicon valley: the anatomy of an entrepreneurial region. Stanford University Press, Stanford

Kloosterman R (2008) Walls and bridges: knowledge spillover between superdutch architectural firms. J Econ Geogr 8(4):545–563

Kochan D (1986) CAM: developments in computer integrated manufacturing. Springer, Berlin

Krugman P (1991) Geography and trade. The MIT Press, Cambridge, MA

Lampel J (2011) Institutional dynamics of project-based creative organizations: Irving Thalberg and the Hollywood studio system. In: Cattani G et al (eds) Project-based organizing and strategic management, vol 28, Advances in strategic management. Emerald, Bingley

Levitt RE et al (2012) Encouraging knowledge-sharing in engineering firms—part I: incentives, disincentives, and the impacts of firm context. Eng Proj Organ J 2(4): 231–239

Levitt RE et al (2013) Encouraging knowledge sharing in engineering firms—part II: game theory analysis and firm strategies. Eng Proj Organ J 3(1): 22–31

Lootsma B (2000) Superdutch: new architecture in the Netherlands. Princeton Architectural Press, New York

Lucertini M, Gasca Millán A, Nicolò F (2004) Technological concepts and mathematical models in the evolution of modern engineering systems: controlling, managing, organizing. Birkhäuser, Basel

Luke R, Begun J, Pointer D (1989) Quasi-firms: strategic interorganizational forms in the health care industry. Acad Manag Rev 14:1–14

Marabelli A (2011) Ma non eravamo i più bravi? Marmomacchine Magazine 218:8

Micelli S (2011) Futuro artigiano: L'innovazione nelle mani degli italiani. Marsilio, Venezia

Naticchia B, Cacciaguerra G (2011) Metodi e strumenti di gestione della costruzione. In: Trento A (ed) Verso un sapere tecnico condiviso nella ricerca sulla progettazione e costruzione dell'edilizia, Atti della Giornata di Studio. Associazione Scientifica Ar.Tec., Roma

Noble DF (1979) Social choice in machine design. In: Zimbalist A (ed) Case studies on the labor process. Monthly Review Press, New York

Noble DF (2011) Forces of production: a social history of industrial automation. Transaction Publishers, New Brunswick

Olexa R (2001) The father of the second industrial revolution. Manuf Eng 127(2)

Orsenigo L, Pammolli F, Riccaboni M (2001) Technological change and network dynamics: lessons from the pharmaceutical industry. Res Policy 30(3):485–508

Ortiz EL (2010) On the impact of philosophical conceptions on mathematical research. The case of Condillac and Babbage. Metatheoria 1(1):65–76

Osservatorio Nazionale Distretti Italiani (2011) Secondo rapporto dell'Osservatorio Nazionale Distretti Italiani. Federazione dei Distretti Italiani, Mestre

Overby A (2011) CNC machining handbook: building, programming, and implementation. McGraw-Hill, New York

Pease W (1952) An automatic machine tool. Sci Am 187(3):101–115

Pérez-Gómez A (1985) Architecture and the crisis of modern science. MIT Press, Cambridge, MA

Phene A, Fladmoe-Lindquist K, Marsh L (2006) Breakthrough innovations in the U.S. biotechnology industry: the effects of technological space and geographic origin. Strateg Manag J 27:369–388

Picon A (1992) Gestes ouvriers, opérations et processus techniques. La vision du travail des encyclopédistes. Recherches sur Diderot et sur l'Encyclopédie 13:131–147

Romozzi F (2010) I vantaggi nello sviluppo delle Aree Leader. Roma

Ross TD (1978) Origins of the APT language for automatically programmed tools. In: Wexelblat RL (ed) History of programming languages I. ACM, New York

Sabel CF (2004) Mondo in bottiglia o finestra sul mondo? Domande aperte sui distretti industriali nello spirito di Sebastiano Brusco. Stato e mercato 70–72(1):143–158

Sinopoli N, Tatano V (2002) Sulle tracce dell'innovazione. Tra tecniche e architettura. FrancoAngeli, Milano

Tatum CB (1989) Organizing to increase innovation in the construction firm. J Constr Eng Manag 115(4):602–617

Taylor JE, Levitt RE (2005) Inter-organizational knowledge flow and innovation diffusion in project-based industries. In: Proceedings of the 38th annual Hawaii international conference on system sciences (HICSS' 05), Waikoloa

Taylor JE, Levitt RE (2007) Innovation alignment and project network dynamics: an integrative model for change. Proj Manag J 38(3):22–35

Valmalette JM (1957) Le dessin technique normalise. Vuibert, Paris

Womack JP, Jones DT, Roos D (1990) The machine that changed the world. Rawson Associates, New York

Zeitlin J (1995) Flexibility and mass production at war. Aircraft manufacture in Britain, the United States, and Germany, 1939–1945. Technol Cult 36(1):46–79

Zeller C (2002) Project teams as means of restructuring research and development in the pharmaceutical industry. Reg Stud 36(3):275–290

Zucker L et al (1996) Collaboration structure and information dilemmas in biotechnology: organizational boundaries as trust production. In: Kramer RM, Tyler TR (eds) Trust in organizations. Sage publication, Thousand Oaks

Chapter 4
Design for Manufacture

Abstract This chapter considers the *digital continuum* between design and manufacture, achieved with the technologies of CAD/CAM, numerical control and prototyping. Although the 'digitalization' of architecture is sometimes seen as a threat to the physical aspects of construction, the opportunity to use a digital model directly in the context of production actually strengthens the traditional links between design and practical craftsmanship. The chapter offers an overview of manufacturing processes currently or potentially relevant to construction. Units of manufacturing processes may be described in sufficiently general terms so as not to be restricted to working with a specific material, a given component or a certain producer of machine tools. Key unit manufacturing processes are identified: (1) mass-change, (2) phase-change, (3) structure-change processes, (4) deformations, (5) consolidation, including (6) rapid prototyping. It concludes by considering integrated manufacturing. Two projects are examined in detail because they have been precursors of the principles of design for assembly: Frank Gehry's "Fish-sculpture" at the Olympic Village, Barcelona (1989–1992), and Norman Foster's Swiss Re Tower in London (1997–2004). The analysis of these projects confirms the recurring importance of the "craft" factor. Computer-aided design and production systems have shifted the emphasis from skills in craftsmanship to competence in the control of the manufacturing machinery. As manufacturing and construction become increasingly automated, more and more skills are required of the designers.

> Much of the material world today, from the simplest consumer products to the most sophisticated airplanes, is created and produced using a process in which design, analysis, representation, fabrication and assembly are becoming a relatively seamless collaborative process that is solely dependent on digital technologies— a *digital continuum* from design to production (Kolaveric 2005).

Branko Kolarevic (2005), William Mitchell (1998) and Antoine Picon (2004) believe innovation in the relationship between design and manufacture to be more significant and relevant than changes in the way of representing the design.

L. Caneparo, *Digital Fabrication in Architecture, Engineering and Construction*,
DOI 10.1007/978-94-007-7137-6_4, © Springer Science+Business Media Dordrecht 2014

> The digital convergence between design processes and production processes offers an important opportunity for a profound transformation of the profession, and thereby of the whole building sector (Kolarevic 2005).

From the possibility of finally reconciling prefabrication and customisation to the promises of robotisation, architecture bears the mark of a rapidly changing context of production. This context might very well lead to a redefinition of the professional identity of the architect, besides modifying the nature of his production (Picon 2010, p. 11).

Kenneth Frampton (1995), in the context of studies of tectonic culture, considers the "digitalisation" of architecture a threat to the physical aspects of construction and architectural technologies. This widespread concern is to be shared if we consider the strongly formal nature of digital design used by several architects. Computer representations such as renderings often tend to neglect the material and technological nature of architecture. Or they confuse it with the restitution of the superficial level of architecture through the procedure of texturing, i.e. applying a photograph over the surface of the design. On the computer screen, unfortunately, shapes seem to be able to float freely, without reference to their context, to the construction or to materiality.

> There is something deeply unsettling in this apparent freedom that seems to question our most fundamental assumptions regarding the nature of the architectural discipline (Picon 2004, p. 114).

The dematerialisation of architectural design did not begin with CAD and rendering. Already in 1983, before the spread of digital technology in architectural design, Gregotti warned in "Casabella" that

> there was the illusion that quotation is a sufficient substitute for the detail as a system of articulation in architectural language, and that an overall "grand conception" can dominate and automatically permeate every aspect of the project and its realization, by the very abstention of the detail, thus polemically underlining the lack of influence of building techniques as an expressive component. Often the outcome of this idea in built terms is an unpleasant sense of an enlarged model, a lack of articulation of the parts at different scales: walls that seem to be made of cut-out cardboard, unfinished windows and openings; in sum, a general relaxing of tension from the drawing to the building.

Certainly the software programs on the market have increased the capacity for reproduction and repetition and have multiplied to the nth degree "the lack of influence of building techniques as an expressive component" criticised by Gregotti (1983). It is thus appropriate to consider the words of Manuel Castells:

> Technology does not determine society; it embodies it. But neither does society determine technological development; it uses it (Castells 1996).

The *digital continuum* between design and manufacture, achieved with the technologies of CAD/CAM, numerical control and prototyping, is antithetic to the dematerialisation of design. The ability to model the design digitally and thus to use this model directly in the context of production creates a synthesis between design and construction, in line with the tradition of the close relationship between design and craftsmanship, between the quality of the design and the rules of the art of crafts.

In the 1950s and 1960s, Italy had illustrious protagonists in Franco Albini, Roberto Gabetti and Aimaro Isola, Carlo Scarpa, or Mario Ridolfi, who interpreted in personal ways the relationship between design, craftsmanship and construction:

> All those craftsmanship potentials which for a long time have been far from the possibility of integration with the design process are being regained (Petrignani 1967, p. 231).

> There is no doubt that at the time of the Bottega d'Erasmo in Turin, we found ourselves looking at a sort of craftsmanship which was slowly disappearing, and in a vision a rather like that of William Morris, we thought that it could be expressed in new ways. And as far as possible, this happened; [...] Ultimately we did not underestimate the fact that so-called 'innovative' technologies, materials and processes might be not abstract entities, detached from the real world, but represented by people, companies, technical systems, workers and colleagues active in our line of work. We must in fact be careful when developing a theory which detaches technology from craftsmen; it is one thing to have a theoretical knowledge of a certain technical possibility, but it is quite another to have the physical human capacity to translate it into a complex task, a scheme traced out by the architectural design. [...] this is why in design it is extremely important to know precisely the character of every firm of craftworkers, indeed of every firm, large or small; understanding its skills and its idiosyncracies (Gabetti and Isola 1995, pp. 84–85).

> Whereas in most sectors of industry, including the neighbouring area of industrial design, the designer asks the producer for advice and collaboration from the earliest ideas stage, with a view to checking that his plans match the technical possibilities of the sector, this synergistic approach is rare in construction, where checking normally happens on the building site against the finished construction design shortly before it is put into practice. Today, however, dialogue with manufacturers from the earliest stages is becoming increasingly necessary, especially for complex forms. In particular the renewed possibilities of industry, offering not just products but a specific and flexible know-how by means of made-to-measure systems of production, thanks to mass-customization or to personalised production, seem to give a firm response to the search for technical solutions suitable to high-performance complex forms (Paoletti 2006).

This also applies to research and teaching;

> Italian higher education departments have placed the term "building production" (produzione edilizia) at the core of the description of their area of work; Florence has "processes and methods of building", others have "building and industrial design" etc.; "building" constitutes the fulcrum of the main reference. That "building" which, whether in terms of the construction market, or in terms of production processes, or in terms of technical innovation, has almost constantly implied research and the profession itself (Del Nord 1999).

As for the relationship between design and industry, Eduardo Vittoria examines how

> intuition, techniques, ability and innovation, that is the capacity to build continuous bridges between the present and the future, between declining products and new products, favour a new concept of knowledge, no longer the property of the traditional owners of knowledge, but also owned by the producers of techniques (Foti 1999).

Eduardo Vittoria's musings highlight the question of innovation, of how to create the "bridges" towards increased know-how. Specifically, he highlights the continuum between design and production actuated by digital technologies. This chapter thus intends to consider questions of building design raised by innovation in digital technologies.

The question of the relationship between production and technology began to be raised in a new way from the second half of the 1700s.[1]

Whilst the unprecedented possibilities of applying techniques to construction and manufacturing were being defined, technical intellectuals, encyclopaedia compilers, scientists, architects and engineers were trying to re-integrate technical skills into the general structure of knowledge. Thus the idea emerged of a science of technics, a technology which would make it possible to bring order to the tumultuous mass of the new skills and competences. What became clear at the time, through the experience of various attempts and approaches, is still the case today: that it has not been possible fully to achieve a general technology which would enable the understanding of the whole set of knowledges useful for designers. The unified system of knowledge was in crisis even before the unprecedented expansion of knowledge which came with the industrial revolution (Picon 1994). There remained the methodological unity around the mathematical sciences and its subsequent developments in information sciences.

4.1   Analysis of the Manufacturing Processes

One of the methods explored during the Enlightenment to organise bodies of knowledge and skills was to organise technics according to their applications. In this way, it is possible to describe digital technologies by means of the manufacturing process, identifying the different machines and classifying them according to their working peculiarities. Because of technological convergence, theorised by Rosenberg (1987),

> the use of machinery in working intended to give a precise form to metal implies a relatively small number of operations (and thus types of machine) [...] moreover, all the machines which perform such operations are found to tackle more or less the same series of technical problems, linked to questions such as the transmission of the driving force (gears, belts, shafts), control devices, or loading mechanisms.[2]

Types of manufacturing processes may be described in sufficiently general terms so as not to be restricted to working with a specific material, a given component or a certain producer of machine tools. Normally, manufacture moves through sequences of operations using a succession of different processes. Fifty years of developing numerical control technologies have produced an extraordinarily large variety of manufacturing processes in different production environments. Using the method once explored by the Encyclopaedia compilers, the working sequence may be broken down into unit manufacturing processes or, more simply, unit processes (Finnie et al. 1995).

---

[1] See Sect. 3.3.1.

[2] See Sect. 2.8.

This chapter aims to give an overview of unit processes currently or potentially relevant to building. Despite the restricted scope of the study, hundreds of unit manufacturing processes are used in the industrial sectors concerned.

To summarize this large variety, unit processes are classified in this chapter according to the manufacturing process with which the configuration or the structure of a material is changed.

In subsequent chapters, on the other hand, production processes will be considered according to the "classical materials" used in building (Torroja 1995); each chapter will start with a mapping summary of manufacturing processes with references to buildings constructed with the technologies being considered.

> We do not intend here to refer to each of the single materials which constitute the building or its components, according to the usual terminology of constructors; we will not speak, for example of cement, but of concrete, or even of reinforced concrete; not of stones, but of buildings in rough or worked stone; not, finally, of bricks, but of brick buildings. That said, we can rearrange the materials as follows: stone, metal, wood and reinforced or prestressed concrete (Torroja 1995).

Finnie (1995) identifies five key unit manufacturing processes: (1) mass-change, (2) phase-change, (3) structure-change processes, (4) deformations, (5) consolidation. Rapid prototyping has been added, as a stand-alone process, because of the particular relevance which it has acquired in the *digital continuum*. The chapter will conclude by considering integrated manufacturing.

4.1.1 Mass-Change Processes

Processing methods refer to the removal of material from a block to make the designed piece. A variety of manufacturing processes belong to this category, for example boring, broaching, shaping, drilling, sawing, punching, milling, grinding, cutting and turning. The basic process consists in removing all material from a block until the remaining piece conforms to the geometry of the design. Depending on the procedure used for removal, the processes may be classified either as chip making processes, in which a tool moves relative to the workpiece, or as processes of subtraction based on lasers, electron beams or on chemical, electrochemical, electrical or thermic erosion.

In summary, the parameters which determine the efficiency and cost of the process come from the characteristics of the material to be worked, the tolerances to be achieved, and the surface finishing of the piece. Other functions which influence cost are operation setup times and working times, investments in the machine, and working costs, which are influenced by tools and energy consumption. Recent developments in machines are towards high-speed operations without loss of quality or of precision.

Manufacturing processes which relate to mass processing include:

- *punching*, in which a tool makes holes of varying depth and diameter in the block;
- *milling*, which uses a mill, a tool with a flat side for cutting that may have cutting edges;

Fig. 4.1 Profile milling (contouring) with circular plates

- *turning* is done on the lathe. The block is round and rotates around its own axis while the tool moves perpendicularly to it, removing successive layers of material;
- *grinding* (polishing), in which abrasives are used to remove thin layers of material and to improve the finishing or precision of the piece made with earlier unit processes;
- *erosion*, in which material is removed from the workpiece by processes based on lasers or electron beams, chemical processes (ECM), electrical processes (EDM), heat or electrochemical processes (Fig. 4.1).

4.1.2 Phase-Change Processes

These processes use the phase changes which different materials undergo either with temperature (in casting a material is brought to a liquid state and is cast into a form) or with pressure (a material is brought to the state of plastic deformation to take the form of the mould). Some of these technologies have ancient origins. Conceptually, the procedure remains similar and is still used in craft and industrial manufacture, as in the cases of casting with sand moulds and lost-wax. The thousand-year history of forging, bending metal and drawing, today controlled by computerised systems, shows the capacity to produce resistant pieces accurately. The development of new materials, reinforced concrete, synthetic polymers, acrylics and resins, engineered and composite materials, plywood, plastic and fibre reinforced materials, requires specific manufacturing processes for making moulds and for forming.

In casting, the raw material is heated until it reaches its liquid or semifluid state. Many materials may be worked in this way. For example, metals, glass, ceramics, artificial stone and some polymers may be heated to the fluid or semifluid stage to be cast or pressure-injected into the cavities of the moulds. Here they are left to cool down until they reach the solid phase when the pieces are taken out.

When the process is applied to the manufacture of metals, it is usually called casting. With polymers, it is instead called moulding. These processes may be executed in the initial stages of the production process in order to make the finished piece or semifinished parts that undergo successive unit processes. The form and the microstructure of the piece determine the efficiency and cost of phase transformations.

Manufacturing processes related to phase-change processes are:

- *casting*: depending on the finished grade of the piece, a distinction is made between die casting performed vertically in single forms for the serial manufacture of pieces and continuous or direct casting by hot-rolling for the production of tubes, bars and sections (these processes may take place early in the flow, the workpieces thus being subject to later process units); permanent casting in moulds which reproduce the negative investment shell of the positive master; the Shaw process and micro-casting in moulds, achieving higher standards of tolerances and finishing.
- *moulding*: polymers are pressure-injected into a mould; the part hardens by compression, thermoforming or reaction with hardeners (Fig. 4.2).

4.1.3   Structure-Change Processes

Several manufacturing processes may induce changes in the structure of many materials. The most common processes are based on heat treatment by heating and cooling in controlled conditions for the specific material, for example hardening, annealing and tempering. Other processes include mechanical treatments to produce plastic deformations and forging. Thermomechanical processes are used to introduce changes that could not be achieved otherwise.

The changes may be confined mainly to the surface of the piece. Preliminary processes consist mainly of cleaning and washing. Surface processing uses heat processes (flaming and annealing); mechanical processes (shot peening and sandblasting) and electrochemical processes (electroplating). A particularly relevant process is the deposition of coatings, which may be metallic or organic. The deposition may be carried out by casting, ion plating, immersion, rolling or spraying, or increasingly commonly by the electroplating processes, anodising, organic coating (paint), glazing or enamelling. Vapours, ions, radio frequency, cathodes and autoclaves are used for the deposition of micro-films, allowing precise control of the width and

Fig. 4.2 Mechanised plant after the operations of casting for the production of casts between 50 and 500 kg, also for small series

the composition of the layer. The form, the chemical properties of the substrate and the finishing of the piece determine the efficiency and cost of the changes to the surface structure. Manufacturing processes related to surface structure-change are:

- *heat treatments* via mechanical processes, electrochemical processes;
- *coatings* deposited chemically, electrically, with ions or mechanically;
- *alloying* through the formation of surface alloys (Figs. 4.3, 4.4, and 4.5).

> Many current devices are quite sophisticated. The sputtering device […] involves molecules being literally drawn from a source and deposited on a substrate in a controlled way. Extremely thin films (some can even be at the nanometre level) can be formed (Addington and Schodek 2005)

4.1.4   Deformation Processes

Deformation processes give the materials a new form. Normally the raw materials, in the semi-finished form of ingots or plates, are plastically deformed by tools which apply forces either of compression, cold rolling, moulding, extrusion, flexing, bending, rolling, pressure-traction, deep drawing or drawing. The process is effected on materials with high deformability either with or without heat. Although cold deformation requires higher forces, it produces parts with better finished surfaces

Fig. 4.3 Plasma processing plant, using partially or totally ionised gas, in which neutral molecules, positive ions and free electrons are present at the same time

Fig. 4.4 Deposition process using the technique of PECVD (Plasma Enhanced Chemical Vapour Deposition), based on the excitation and ionisation of silicon (SiOx), it creates a nanostructured polymer film. The polymer is deposited on the surface of the wood and becomes all in one with it. The film has a chemical composition similar to quartz (with a high elasticity), and makes the surface of the wood water-repellent, protects it from chemicals and organic solvents and resists UV light. After ionic deposition, only one coat of varnish is required

and precision of size. The pieces obtained by deformation generally have the advantage of saving material and work compared with mass-change processes. They have better mechanical characteristics in that the fibres of the material are stretched, but not broken.

Fig. 4.5 Wood samples after 6 years of exposure to the elements: on the *left* treated with PECVD, on the *right* not treated

Successive combinations of these unit processes allow the manufacture of complex parts from simple semi-worked elements with high productivity. For this reason, deformation processes are commonly used in mass production.

The main deformation processes used in manufacturing are:

- *lamination*: the desired thickness is obtained by stretching and forming the piece, deformed by the pressure of passing through two or more rotating rollers;
- *moulding*: the piece is forced to take on the shape of the die through pressure and bending;
- *forging*: the piece is beaten with a trip-hammer or pressed to assume the desired shape;
- *extrusion*: the piece is forced to pass through a draw-plate which moulds it to its shape (Figs. 4.6, 4.7, 4.8, and 4.9).

4.1.5   Joining and Consolidation Processes

Joining processes consist of assembling different parts into a single piece, in order to achieve the desired geometry, structure or property. These processes have the aim of permanently or temporarily joining two or more parts, creating a bond, based on various procedures: mechanical, chemical or heat processes. The interaction between the material and the energy which creates the bond is a key element of the

Fig. 4.6 Processes of slicing and moulding with an electronically-fed press

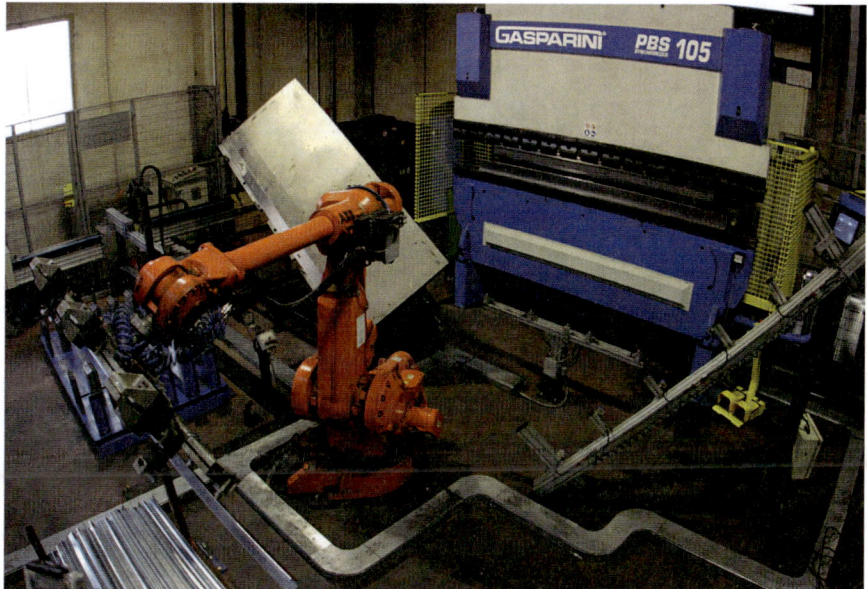

Fig. 4.7 Hydraulic bending press controlled by CNC for bending semi-worked metal pieces. The press is controlled by a robot, which executes the manipulation, loading and unloading of the pieces

Fig. 4.8 Panelling area at OMR Snc, which performs all the operations involved in bending a punched and cut piece so as to transform it into a finished component. The loading and all bending operations with the necessary rotations and relative positioning are automatic. Finally the component is removed from the bending centre by the robot which manages the palletisation

Fig. 4.9 Robotised work areas for pressing-bending and for panelling are useful in making medium-large series: single sheet doors from OMR Snc

process. In some processes, the consolidating energy improves the structure or the properties of the material and is an integral part of the process. For example, in sinter-forging (powder preform forging), the mechanical energy consolidates the powder, gives shape and size to the piece and at the same time improves the microstructure of the material, obtaining better mechanical efficiencies than the material used. In other cases, the energy used for joining is harmful to the structure or the properties of the product. In fusion welding, heat creates the bond between the objects, but it can also alter the area of the joint, inducing damaging distortions and residual stresses. Welding is a process of permanent joining which may be achieved with several procedures depending on the characteristics of the material or materials, the thickness of the area of the joint and the mechanical requirements.

Sintering is a process of powder consolidation through heat processes or a combination of heat and mechanical processes. The procedure consists of compacting and heating the particles in a mould so that the thermal energy creates strong bonds between them. The powders used are of metal, ceramics, composites or a mixture of these. Principal advantages are the high level of precision of the details and finishing, even in complex forms; precise control over microstructural characteristics; extremely good mechanical performance and a high production speed. On the other hand, it requires a specific design, which is impossible when making large pieces and in the case of production in large series.

Manufacturing processes related to joining and consolidation include:

- *welding*: two parts are joined via progressive fusion of the joint. Sources of heat used are gas combustion, electric arc, plasma or laser. It is achieved by adding either the same material as the parts or different material.
- *brazing*: two parts are joined by means of a material in liquid state with a melting temperature significantly lower than that of the materials to be joined which, for this reason, do not assist the join by melting.
- *sintering*: powders are pressed into a mould to give them shape and are then consolidated with heat processing (Figs. 4.10, 4.11, 4.12, and 4.13).

4.1.6   Rapid Prototyping Processes

Processes of rapid prototyping (RP) make it possible to achieve, within hours, objects with unlimitedly complex geometries from the CAD model (Figs. 4.14, 4.15, 4.16, 4.17, 4.18, 4.19, 4.20, 4.21, 4.22, and 4.23). The principal applications of rapid prototyping are in the creation of:

- conceptual models of a piece being designed with the aim of evaluating the aspect and the dimensional characteristics of the design;
- prototypes for validation;
- technical prototypes;
- positive masters, made of wax, photopolymers or composite materials for moulds, constituting the sacrificial mould inside the chalk, sand or clay shells;
- masters for use in moulding with plastic materials;

Fig. 4.10 Bead sintering process for the production of diamond thread, used in cutting stones

Fig. 4.11 Mixing and pressing of the beads

Fig. 4.12 Sintering furnace

Fig. 4.13 The finished beads, ready for the assembly of the diamond wire

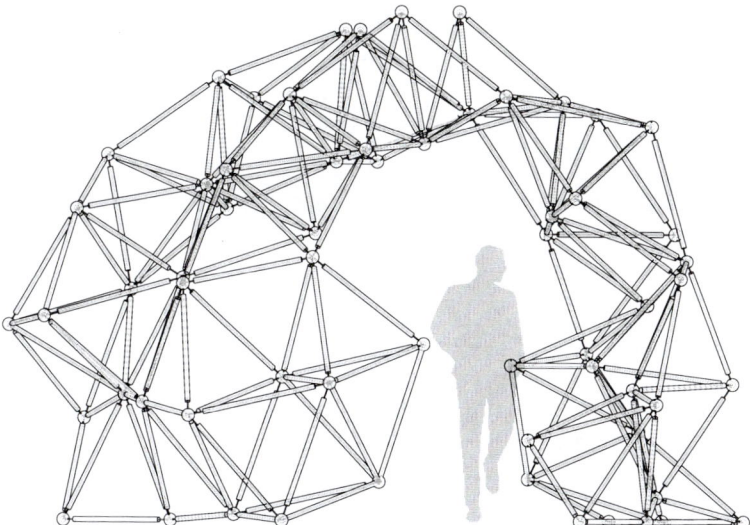

Fig. 4.14 View of the Distort Pavilion, an educational project at the UTS School of Architecture, Sydney, Australia

- a functional prototype, useful for testing the mechanical or structural properties of the piece or for checking the coupling;
- finished parts, single pieces or in very small series.

Given the difficulty of classifying processes which are rapidly evolving (Kai and Fai 2010; Gatto and Iuliano 1998), relevant processes of rapid prototyping function as follows:

- *by consolidation*: a laser beam or UV beam solidifies the surface or small part of a thin layer of fluid and successive layers are deposited one by one as each time the support which holds them up is moved down by the height of the layer;

Fig. 4.15   The erected pavilion with the group which participated in the project

- *by superimposition*: successive layers of papery material are cut by laser and stuck together; thermoplastic materials, ABS, elastomers, wax, or polycarbonates are deposited to build the layer;
- *by sintering*: a laser beam sinters particles of metal, thermoplastic particles or sand until they form the piece.

4.1.7   Integrated Processes

Manufacturing a piece or a component often needs a succession of unit processes in order to make up the production flow. Each unit process carries out a single manufacturing process on the workpiece in a sequence in which the workpiece entering one unit process is the outcome of the preceding process. The finished piece is the product of this sequence and is the sum of the single manufacturing units: the quality and the properties of the piece may be quantified by adding and subtracting the contributions of each unit process.

In order to draw the most complete benefit from the possibilities of a production sequence, it is necessary that the complex process of integration of all the single unit

Figs. 4.16 and 4.17   Prototyping of the structural nodes

Figs. 4.18, 4.19, and 4.20 D-Shape is a rapid prototyping system for making large stone-like objects on the scale of buildings. Successive layers of sand 13⁄64 in (5 mm) thick are deposited. By means of an inorganic binder, the sand sets in a microcrystalline structure, making a stone-like conglomerate with high resistance to traction

Figs. 4.21, 4.22, and 4.23 Manufacturing process for the Radiolaria pavilion, a project by architect Andrea Morgante of Shiro Studio. The shape is inspired by the amoeboid protozoa of the same name. The stages of prototyping: moulding of the monolithic structure; removal of inert materials that were not consolidated and which will be reused; the finished pavilion after finishing and polishing

processes be accurately designed and planned. Computer Integrated Manufacturing (CIM) organises and manages the complete production sequence through the working of the system in all its stages, from the design, engineering, manufacture and quality control to the planning of production relating to marketing (Groover 2001).

In the building industry, the integrated management of the manufacturing processes is accomplished at various levels. The characteristics of semi-finished products are particularly appropriate to integrated manufacturing processes. Many industries use integrated processes in the manufacturing of semi-finished products. In several cases, they are made up of complex products which may assemble different materials, parts or functions. In such cases, integration has been followed through the production process and the overall organisation of the construction sector, from the design to the building site.

Building systems and components are made in integrated plants when sufficient volumes of production are required. Whole process integration has often required the planning of a system for support and maintenance of the specific product within the framework of the overall organisation. In this system, the capacity for commercialisation cannot be detached from support to all phases of design, engineering, coordination and construction. It is possible to set up a relationship between innovative possibilities of components or systems and the ability to introduce and manage integrated processes in the specific field of the construction industry.

The construction sector is characterised by a combination of circumstances,

the uniqueness of the architectural design connected to the peculiarities of the site (Nardi 1992),

and

the role of the traditional company, often based more on land revenue than production issues (Ibid.),

come together to create the one-off and ever-changing nature of the construction
site. It is a combination of factors which has hindered and still hinders the spread of
integrated processes in construction and of innovation in general.[3] Industrial automation has allowed highly integrated plants to be made, resourced by pioneering
researches and experiments in production and automation (Figs. 4.24 and 4.25).[4]

Currently the achievement of integrated processes is limited to a few pilot
projects. In Japan, where the prestige which comes from technological innovation
is important and labour costs are particularly high, several experimental CIM
methods in building have been produced and are in process (Miyatake and Kangari
1993; Yamazaki and Maeda 1998; Yamazaki 2004).

Shimizu Manufacturing system by Advanced Robotics Technology (SMART) is
one of the most advanced pilot projects. The construction method is based on a
structure which reproduces the basic floorplan, where several robots work on the

---

[3] See Sect. 3.4.2.

[4] See Chap. 2.

Fig. 4.25 The platform is lifted from one storey to the next by four jacks

movement of materials (e.g. cranes, trolley hoists); construction (e.g. welding, cutting, finishing); inspection and checking. When the construction of one storey is finished, the structure is pushed or lifted to the floor above, where the construction process starts again. The basic idea is to create a template, the platform, program the robots to build the floor and then repeat the process for any number of floors.

The platform has a covering to protect it from the elements and to allow work to continue without interruptions irrespective of weather conditions. In the construction stage, the building seems to be wearing a "top hat".

The headquarters of two banks have been built with total automatisation of the building site; one in Nagoya in 1993 and one after that in Yokohama. In these two building sites, robots have reduced manual work by 30 %.

The construction of each building took 6 weeks. The platform on which the robots work weighs 1,300 tons and is lifted from one storey to the next by four jacks (Fig. 4.26).

When one floor is completed, the platform is raised up to the next. The lower part of the platform holds overhead cranes used for moving materials, which makes the building site look like a factory.

Fig. 4.26 Shimizu Corp., the building site resembles a factory

The whole construction process is supervised by computers in a control cabin.

Although workers are still present on the building site, their jobs are limited to maintenance and checking. The reduction of the labour force required the ad hoc design of the components. Prefabricated pillars and floors use self-guiding joints to facilitate automatic positioning. Laser-guided systems are used to direct the automatic positioning and subsequent joining by welding robots (Figs. 4.27 and 4.28).

The building site has a high level of integration in automation: raising, positioning and fixing the structure, the floors and the internal and external walls, as well as the installation of prefabricated units. The automation of the processes is achieved starting with the designing and planning and then making the components. At the moment, programming the robotised system is time-consuming and costly, and consequently the building type tends to be a repetition of the basic floorplan.

Up to now, experiences of using CIM in construction have shown that higher levels of process integration have been hampered by several factors, among them limited levels of flexibility in type and structure and difficulties of integration and cost within one or between different sectors, including that of machine tools for the building site, mostly with a specific use and purpose. This gives rise to the need to design and to produce all the machines and to integrate them within the system. This consequently requires a very high investment, which can only be amortised by the building activity of the company financing the innovation.

Fig. 4.27 The whole construction process is supervised from a control cabin

Fig. 4.28 Robotised system for laying floors

4.2   Analysis of the Shape and Specifications of the Piece

Analysis which starts with the desired result (the finished piece or component) is probably more appropriate to the requirements of the designer because it analyses the shape and specifications of the piece with regard to the manufacturing process. The outcome of this analysis is not the identification of machine tools, as it is instead the investigation of manufacturing processes, that is, the breaking down of the basic needs of the working processes and their relationship with the desired outcome, namely the specifications of the piece to be produced. Issues may emerge from this analysis which are specific to a manufacturing process or to a sequence of processes: important indications for the design may also emerge, for example suggestions for possible changes in the piece because of significant production advantages.

The specifications of a piece may be varied and unique. Those with broad applicability may be sorted in requirements for the batch size, the tolerance, the strength, the ease of assembly and the shape.

The size of the batch may vary from a single piece to millions. If a batch of only few components with strength properties is needed, it can be made by machining or casting. Casting needs a costly positive master of wood or wax. This generates the negative investment shell into the sand mould before the melted metal is poured in. The positive master can be made either with a CNC machine or with rapid prototyping. Finally, the piece may be made directly in rapid prototyping by sintering.

Criteria for choosing among the different processes depend on the shape of the piece and the strength required. Lower strength is usually provided by sintering in rapid prototyping. Greater strength is obtained by casting, and it is further increased with chip-making processes and by forging. For complex shapes, particularly if they are curved, casting with a mould made by prototyping is more viable. Whilst for very small batches, less than ten pieces or so, CNC machining is economic in the re-use of the program, for larger lots the cost of making a mould which is not lost is a deciding factor.

Dimensional tolerances of the pieces of the order of 50 μm require costly processes of machining, grinding and polishing. It is a good idea for the designer to consider building and assembly procedures which do not require such finishing costs. Fidelity in the order of thousandths of a millimetre between the CAD model and the manufactured part made cost approximately a tenth of tolerances in the order of 100 μm[5]. A wide range of factors influence tolerances, such as the properties of the specific alloy, of the single operations in the production process and of external factors relating to humidity, friction or dirt.

The shape of the piece contributes to the selection of the processes to be preferred over others. For example, wide thin sections are more suited to sheet working. Uneven sections are more easily made by casting, forging or milling. Cylindrical shapes indicate turning. In general, complicated geometries, with small details, require longer and more costly operations.

---

[5] 0.100 mm.

Methods of assembly are relevant in defining the specifications of the piece. For example the costs of welding, bolting and rivetting are high and may bring problematic factors into the construction, particularly when they take place on site. The manufacturing of a component or system may benefit from the design of the parts conforming to uniformity and simplicity criteria of assembly. The automobile electronics industry has developed and tested systems aimed at design for assembly[6]: modules for the major mechanical CAD programs allow the evaluation of the simplicity of assembly for a component on the basis of type and number of the joints and of the shape of the parts (Stackpole 2010).

In the Architecture, Engineering and Construction industry, various projects have been precursors of the principles of design for assembly. For example Frank Gehry's "Fish-sculpture" at the Olympic Village, Barcelona (1989–1992), and Norman Foster's Swiss Re Tower in London (1997–2004).

4.3   Frank Gehry Partners, Fish Sculpture at the Olympic Village in Barcelona

Frank Gehry designed a structure 180 ft (55 m) long by 115 ft (35 m) high. In the gigantic sculptural form, Gehry could freely express his own figurative research (Figs. 4.29, 4.30, 4.31, 4.32, 4.33, 4.34, 4.35, and 4.36). In successive works, the poetry has evolved into the abstract since the "Dancing House" in Prague (1994–1996). The Fish offered the opportunity and the definite challenge of construction of forms on the scale of buildings without having to handle the restrictions and constraints of a habitable system. In Gehry's studio, James Glymph began to investigate computer-aided design systems for modelling plastic forms and resolving the models into manufacturable sub-models. At first the Fish was modelled with Alias (Maya) in collaboration with Bill Mitchell at MIT, "but that is as far as that system could be used for production."[7] Working in Gehry's office was Rick Smith, hired in July 1991 for modelling the Walt Disney Concert Hall.

Rick Smith owns an assisted-design consultancy company, "C-cubed Virtual Architecture." The three Cs stands for Certified CIM[8] Consulting. Rick Smith has passed a CIM certification process developed by IBM, hence the company's name C-cubed. The company consults in the aerospace industry, with the CAD-CAM system *Computer Aided Three Dimensional Interactive Application* (CATIA), developed by Dassault Aviation. CATIA was born to support the design and production of the Mirage fighter, one of the first aeroplanes to be intensively designed and produced using a three-dimensional CAD-CAM system. CATIA was developed by Dassualt in collaboration with IBM. Lockheed Corporation had developed a

---

[6] See Sect. 3.3.

[7] Interview with Rick Smith, September 2012.

[8] See Sect. 4.1.7.

Fig. 4.29 Frank Gehry's Fish Sculpture at the Olympic Village in Barcelona

Fig. 4.30 Gehry's physical model

Fig. 4.31 CATIA master model of the main structure of the sculpted fish shape

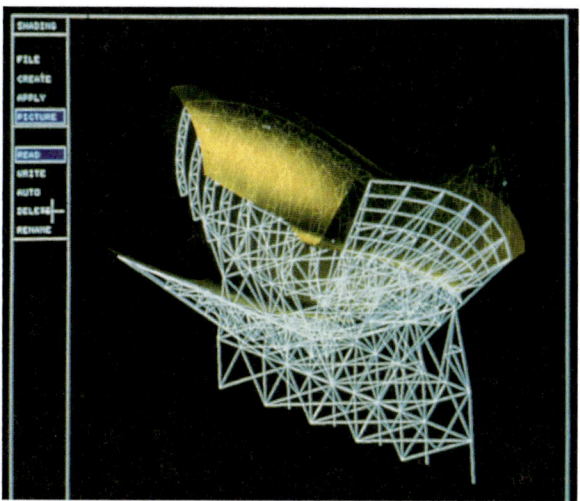

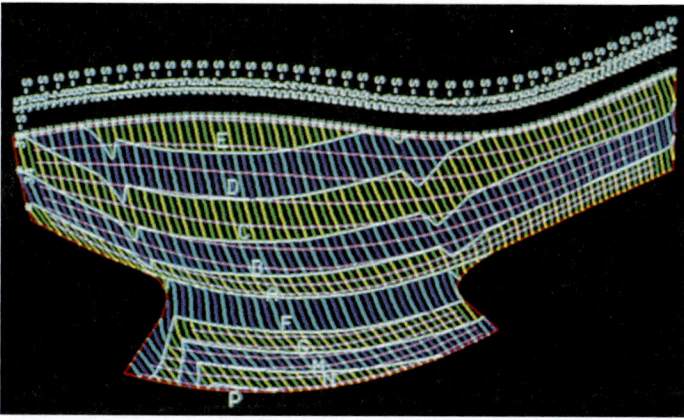

Fig. 4.32 CATIA model diagramming individual secondary structural pipe layout for bending fabrication

2D drafting and CNC system called CADAM (Computer-aided Design and Manufacturing). Dassualt began using CADAM in the 1970s to build their aircraft. Dassualt took the 2D platform and developed a 3D modelling system. IBM implemented the computer systems and interfaces for the software to run on. In 1981, IBM and Dassault began marketing the CATIA system.

Rick Smith started working on the Fish in September, measuring Gehry's physical model and building the 3D model from the measurements. At the time, there was no digitizing equipment in Gehry's office. "Computer digitizing equipment was just beginning to be developed and we began researching the various systems in 1992.

Fig. 4.33 CATIA model
of the structural steel pipes
with connections to primary
structural steel

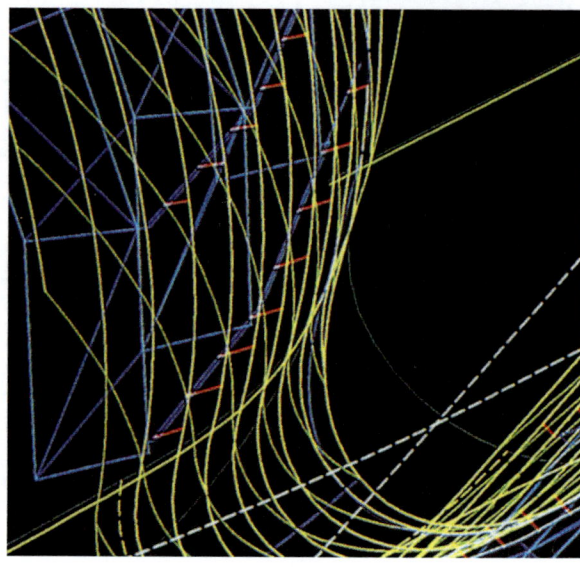

Fig. 4.34 Permasteelisa was
given the job of construction
design and developed a
complete construction
methodology in which
the skin shapes the inside.
The skin defines the
secondary and primary
structures

Fig. 4.35 A database keeps a record of each piece's unique identification code, its place in the gantry and its progress

After a number of iterations in modeling the shape of the fish in CATIA Frank approved the shape."[9]

Smith modelled the Fish, both the surface and the internal structure, according to a geometric model which traced and transmitted to the interior the form of the skin.

Permasteelisa was charged with the construction design and with developing the geometric method into a complete construction methodology which could shape the inside from the outside.

From the approved digital model, a number of 3D models were developed to match different design requirements. Skidmore, Owings & Merrill (SOM) were charged with the design of the primary structure. They used Architecture Engineering System (AES), which SOM had been developing since 1980. AES geometric primitives are line diagrams, so the Fish 3D shape at the surface was translated "to line segments that were acceptable to diagram the stick centerlines of the primary structure."[10]

---

[9] Interview with Rick Smith, September 2012.
[10] Ibid.

Fig. 4.36 Rather than the
blueprints, it is the database
which manages and organises
the progress of the project: it
allows the tracking of each
single part from manufacture
to the building site
and checking the work
progress status

SOM's line diagram of the primary structure was imported back into CATIA and
Rick Smith used it to model the secondary structure. From December onwards he
worked on designing the parallelogram panel cladding. The cladding pattern drove
the design of the secondary structure and subsequently the attachment points to the
primary structure.

The panelling was a unique process. There was only one panel design. I think it was 1
meter square. It was made up of four strips of a few centimeter wide stainless steel, running
horizontally and vertically. The strips were attached to each other with one rivet. This
allowed the panel to shift as a parallelogram. So as I mapped the panels onto the surface
the panel would adjust from a square shape into a diamond shape. I developed a technique
in the computer to map the panels three dimensionally along the surface. It was like
painting yourself into a corner. Depending on certain design criteria I obtained three
different panel schemes on the surface over the course of Christmas vacation. After
Christmas I presented them to Frank. He chose one of the schemes. From that pattern
scheme I developed the three dimensionally curving structural pipes we called the strakes.
I then took all the models to Permasteelisa in Italy. From those models we detailed the

model more fully in every way necessary for fabrication. Detailing every joint, connection, panel, structure, etc. This way we had the details of all the parts to fabricate and dimensions for erection on the job site. We said, "We virtually built the project before the project was built." We produced drawings from CATIA but most importantly the 3D model was used for fabrication of the complex parts. For example, we had to develop a 3D jig concept to fabricate each individual 3D curved strakes. On the job site the model was used directly to extract each connection dimension. A spreadsheet was used by the erector with each connection location and dimension length.[11]

For this procedure, an idiomatic expression was borrowed from the automobile industry, "skin in", according to which the 1:1 scale model of the bodywork acts as a reference for the development of components and internal subsystems for the vehicle (Saggio 2001, pp. 6–7).

Permasteelisa acquired CATIA and in February Smith arrived at the company's headquarters in Vittorio Veneto, Italy, to work in direct contact with the design office, particularly with engineer Marzio Perin in charge of the use of CATIA in the fabrication. Together they developed some pioneering techniques for translating the 3D model into all the necessary detailed construction entities, to bend the 3D curved structure and panel fabrication mutually. They used concepts that Smith had learned in the aerospace industry, for example using 3D jigs to bend the 3D structural tubes.

At the Permasteelisa plant, a full scale mockup of about 1/8th portion was built. Permasteelisa used an airplane engine with a propeller mounted to blast a 150 kph wind against the mockup for testing the panel attachment method.

On site, construction was achieved by a metallic scaffolding of tubes covered by a bronze colour-dipped stainless steel mesh.[12]

CATIA integrates CAM modules for direct interface with the manufacturing processes; in this way, each of the thousands of connecting elements comprising the structure can be produced. A database keeps a record of each piece's unique identification code, its place in the gantry and its progress (Fig. 4.35). Rather than blueprints, it is the database which manages and organises the progress of the project: it allows the tracking of each single part from manufacture to building site and the checking of progress. During assembly in Barcelona, only two joints ended up misaligned by a few millimetres (Novitski 1992).

Our initial experiments in this were astoundingly successful. It shocked us all. We did our first paperless process in 1991 on a Fish sculpture in Barcelona. It created a wonderful collaborative environment between us and the fabricators and builders who executed the work. The project was finished perfectly to what Gehry was looking for, ahead of schedule, on budget and everyone enjoyed the process, which was one of the most important things (Day 2004).

The Fish project marks the beginning of the long collaboration between Permasteelisa and Gehry Partners.

---

[11] Interview with Rick Smith, September 2012.

[12] Made by Steel Color S.p.A. The choice of steel was dictated by the proximity to the marine environment and its corrosive elements. After 20 years the material is unaltered.

4.4   Foster + Partners, Swiss Re Tower in London

Norman Foster designed a 50-storey tower 590 ft (180 m) high for Swiss Re in the City of London (4.37, 4.38, 4.39, and 4.40).

The needs of the project were numerous and varied: a floor space of 500,000 sq. ft (46,450 m²); plenty of public spaces within the new building; high environmental sustainability for the overall system; accessibility by public transport (no parking space); improved building microclimate; predominantly natural lighting; and reduced energy consumption on air-conditioning.

Foster + Partners have succeeded in interpreting the requirements in a magnificent organic form,[13] a successful synthesis of the requirements. The circular plan on the entrance floor is 131 ft (40 m) in diameter, which expands to 187 ft (57 m) in the central section. The reduced footprint leaves room for a vast pedestrian open space, on which several activities and services have begun.

The basic floor plan is organised around a central area which hosts services, stairs, lifts and the pillars which bear the vertical loads. We will call this central area, the "hub" of a metaphorical wheel, from which six "spokes" go out, the usable floor space of each level. Each spoke is separated from the next by a triangular area, about 20° wide and vertically open, which creates light wells for almost the entire height of the building, up to the 32nd floor. The purpose of these wells is to aid ventilation and natural lighting.

The building is generated by rotation around the hub: each floor is rotated by 5° relative to the one below. The circular movement between floors makes the light wells go round the building: shown by the grey-blue windows of the façade which give the building its recognisable spiral development. The façade also shows the distribution of the forces in which the horizontal loads are discharged in the nodes of the networked outer structure (Fig. 4.41).

Contributing to the statics of the building are the central core and the perimeter grid of diagonally interlocking steel columns, known as the *diagrid*.

The diagrid relieves the lateral loading from the central core, which contributes to the load bearing capacity. The horizontal loads from the floors are transmitted to the diagrid through planar hoops connecting the slabs to the nodes joining the circumference columns.

The diagrid provides the dynamics of the structure with improved conditions of balance in the case of asymmetric or lateral loads, contributing significantly to the rigidity of the system: wind pressure at a height of 590 ft (180 m) causes elastic movement of 2 in (50 mm) (Munro 2004, Fig. 4.42).

> The diagonal networked structure of the Swiss Re building was achieved with A-shaped frames developed over two floors, composed of two diagonal columns of tubular steel 508 mm in diameter and 32–40 mm thick; a cross tie beam of tubular steel sections in square cross-section of 250 mm by 250 mm; and a steel node. The frames are connected to the nucleus [the hub – Author's note] with the help of radial beams in rolled steel. The façade takes on the diagonal geometry of a networked mesh, reflecting it in the triangular and diamond-shaped backgrounds of the window panes (Compagno 2003).

---

[13] Londoners have, perhaps affectionately, christened the building the "gherkin".

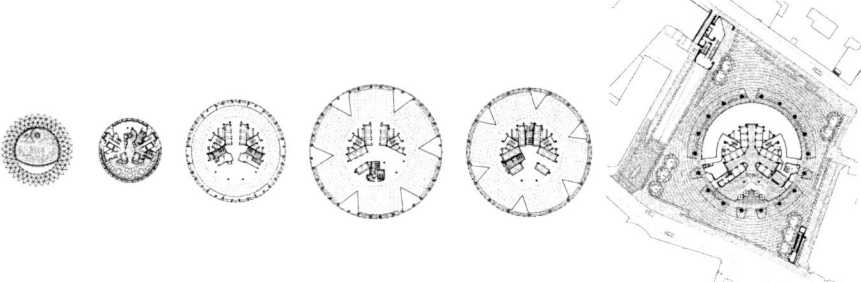

Fig. 4.37 Foster+Partners. Swiss Re Tower plan of the 50th, 39th, 33rd, 21st and 6th floors and the entrance with the arrangement of the public spaces

Fig. 4.38 The tower has a diameter of 131 ft (40 m) on the ground floor, and expands to 187 ft (57 m) in the central section

Fig. 4.39 The tower is
590 ft (180 m) high on
50 floors

Fig. 4.40 Foster + Partners interpret the needs of the commissioners in a magnificent organic form
which has become a tourist attraction and a conspicuous part of the London skyline

Fig. 4.41 Cross-section view of the tower

Fig. 4.42 Foster's sketch shows the process of generating the building around the hub: each floor is rotated by 5° relative to the one below. The circular movement between floors makes the light wells go round the building

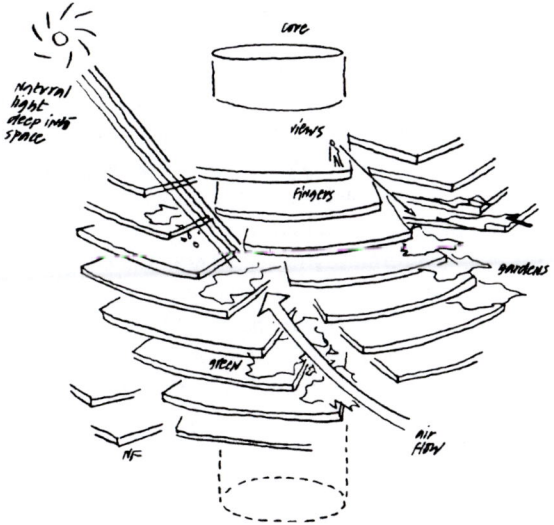

Fig. 4.43 The building site
shows the characteristic
diagonal geometry of the
networked structure, achieved
with A-shaped frames
connected to the horizontal
hoops of tubular steel sections
which encircle each floor

The node must guarantee stability across three different planes: the horizontal
hoop of tie rods; the upper columns and the lower columns. Because of the varying
width of the building, the geometry of the structural nodes changes with each storey.
Ove Arup project team considered and compared alternative design strategies for
manufacturing and assembling the parts. One possible strategy included designing
and making single nodes, columns and hoops to fit the needs of each individual level
of the structure. Another strategy aimed at standardising the node so that it would be
suitable for a predetermined variety of needs and performances while managing
differences in size and structure by varying the columns and hoops. The latter possibility
was adopted, as it yields consistent advantages in manufacture and in assembly on site
since it is uniform for the sake of simplification (Munro 2004, Figs. 4.43 and 4.44).

The design of the node is the result of a close collaboration with the contracting
company, the joint venture between Victor Buyck Steel Construction and Hollandia
BV. The company developed its own innovative system of anchoring the node to the
metalwork of the floor: a counter-plate, fixed to the overhang of the floor, is bolted
to a corresponding plate attached to the node. This solution guarantees the stability
of the horizontal hoop, the transmission of the horizontal loads on the level of the

Fig. 4.44 The building site of the Swiss Re Tower; in the foreground are the node-linking hoops, columns, and the ties for the fixtures. Each piece is numbered sequentially so that it can be immediately identified and checked

floor and the possibility of adjusting the position of the node during assembly to adapt it to the tolerances of the construction project. Self-adjustment is achieved by fixing eccentric bolts, which is feasible because of the high precision of the structure as a whole. Self-adjustment simplifies the assembly process in that it allows the horizontal hoop of tie rods to be closed on each floor without having to find solutions for more difficult and complex adjustments, such as fixing running guides in the shop floor or pre-stressing the bolts.

So that the tolerances of the node conform to those of the structural system, a maximum margin of a tenth of a millimetre between planned and actual dimensions was accepted in its manufacture. Particular attention was given to making the plates which link nodes and columns, which are made by milling.

The development of the diagrid structure is highlighted by its correspondence with the triangular pattern of the glazed façade.

The triangular tessellation is also the most geometrically simple and efficient way of dealing with the variations in curvature along the façade, guaranteeing the flatness of each window, constructed by putting two isosceles triangles together in a diamond-shape. In fact, the only curved element of the façade is at the top of the building (Fig. 4.45).

Although the design aims for simplification, each level of the glass façade has a different shape.

Fig. 4.45 The triangular
tessellation of the reticular
structure of the façade
coincides with the windows
constructed by juxtaposing
two isosceles triangles in a
diamond-shape

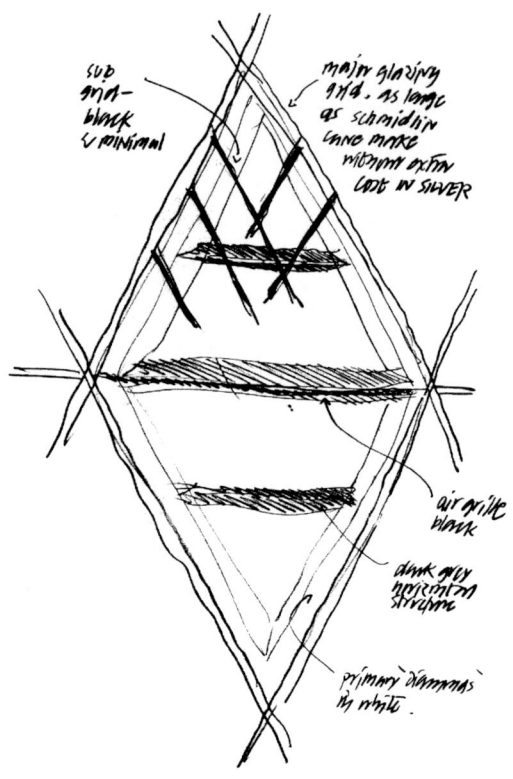

The differences in shape and size of the pieces necessitated a specific method of production,
loading and assembly which was only possible with a high level of computerisation
(Compagno 2003).

The facade frames are made of thermal break aluminium profile, powder coated;
after hardening in a furnace, a strong and tough coating is formed. Particular attention is paid to the system of anchoring the frames to the façade by a standard join
which sums several important factors. A single join links the vertices in two adjacent
frames, which helps to reduce both the number of joins and the number of links for
each panel. Any onsite adjustment and regulation of the façade's join to the structure
are resolved with only two bolts that control the rail guide. During assembly, the
frames are lowered from above and fixed in a single operation.

In conclusion, the close examination of the two projects, those by Gehry and Foster
respectively, confirms that design for manufacture depends on the iteration between
several factors. Of recurring importance is the "craft" factor, which comes from the
experience and understanding of the designer, or more often the design team.
Computer-aided design and production systems have shifted the emphasis from skills
in craftsmanship to competence in the use of the machinery which handles the manufacturing. The higher the level of automation in manufacturing and construction, the
higher the competencies required of the designers (Compagno 2003).

References

Addington DM, Schodek DL (2005) Smart materials and new technologies: for the architecture and design professions. Architectural Press, Oxford

Castells M (1996) The rise of the network society. Blackwell Publishers, Malden

Compagno A (2003) Grattacieli, Swiss Re + Westhafen. Detail 7/8:6

Day M (2004) Architect Frank Gehry finds cad a boon to art and business. CAD Digest 23:64–78

Del Nord R (1999) Saluto del Direttore del Dipartimento di Processi e Metodi della Produzione Edilizia. In: Buccolieri CC, Giallocosta G (eds) Progetto e produzione nello scenario edilizio contemporaneo: questioni e contributi. Alinea, Firenze, Atti del Convegno nazionale Costruire verso il 3° millennio, dedicato a Mauro Maccolini

Finnie I et al (1995) Unit manufacturing processes. National Academy Press, Washington, DC

Foti G (1999) La trasformazione del costruire. In: Buccolieri CC, Giallocosta G (eds) Progetto e produzione nello scenario edilizio contemporaneo: questioni e contributi. Alinea, Firenze, Atti del Convegno nazionale Costruire verso il 3° millennio, dedicato a Mauro Maccolini

Frampton K (1995) Studies in tectonic culture. MIT Press, Cambridge, MA

Gabetti R, Isola A (1995) L'architettura del colloquio. In: Faroldi E, Vettori MP (eds) Dialoghi di architettura. Alinea, Firenze, Franco Albini, BBPR, Lodovico B. Belgioioso, Guido Canella, Aurelio Cortesi, Ignazio Gardella, Vittorio Gregotti, Vico Magistretti, Enrico Mantero, Paolo Portoghesi, Aldo Rossi, Giuseppe Terragni, Vittoriano Viganò

Gatto A, Iuliano L (1998) Prototipazione rapida. La tecnologia per la competizione globale. Tecniche nuove, Milano

Gregotti V (1983) The exercise of detailing. Casabella 492:10–11

Groover MP (2001) Automation, production systems, and computer-integrated manufacturing. Prentice Hall, Upper Saddle River

Kai CC, Fai LK (2010) Rapid prototyping: principles and applications in manufacturing. Wiley, Singapore

Kolarevic B (2005) Architecture in the digital age: design and manufacturing. Taylor & Francis, New York

Mitchell WJ (1998) Antitectonics. The poetics of virtuality. In: Beckmann J (ed) The virtual dimension. Princeton Architectural Press, New York

Miyatake Y, Kangari R (1993) Experiencing computer integrated construction. J Constr Eng Manag ASCE 2(119):307–323

Munro D (2004) Swiss Re's building London. Nyheter Stålbyggnad 3:37–42

Nardi G (1992) Gli orizzonti dell'industria delle costruzioni. L'arca 66:90–91

Novitski BJ (1992) Gehry forges new computer links. Architecture 8:105–111

Paoletti I (2006) Costruire le forme complesse. Innovazione, industrializzazione e trasferimento per il progetto di architettura. Clup, Milano

Petrignani M (1967) Disegno e progettazione. Dedalo, Bari

Picon A (1994) Les rapports entre sciences et techniques dans l'organisation du Savoir. Revue de synthèse 1–2(4):103–120

Picon A (2004) Architecture and the virtual towards a new materiality. Praxis j writ build 6:114–121

Picon A (2010) Digital culture in architecture: an introduction for the design professions. Birkhäuser, Basel/Boston/Berlin, p 11

Rosenberg N (1987) Le vie della tecnologia. Rosenberg & Sellier, Torino [Italian edition of Perspectives on Technology]

Saggio A (2001) Preface: flying carpets. In: Lindsey B (ed) Digital Gehry: material resistance, digital construction. Birkhäuser, Basel/Boston/Berlin

Stackpole B (2010) DFMA takes a back-to-basics product simplification strategy to cut costs. Des news mag 9

Torroja E (1995) La concezione strutturale. Logica ed intuito nella ideazione delle forme. Italian edition edited by Franco Levi, Città Studi, Milano. English edition: (1958) Philosophy of structures. University of California Press, Berkeley

Yamazaki Y (2004) Future innovative construction technologies. Directions and strategies to innovate construction industry. In: Sung-Keun Kim (2003) Construction automation. In Chen WF, Richard Liew JY (eds) (2003) The civil engineering handbook. CRC Press, Boca Raton

Yamazaki Y, Maeda J (1998) The SMART system: an integrated application of automation and information technology in production process. Comput Ind 1(35):87–99

Chapter 5
Digital Manufacture of Metals

Abstract More than two centuries of research and development have produced a vast choice of metallic materials and manufacturing processes, useful in giving form and structure to architecture. This chapter aims to consider the state of manufacturing technologies in the light of possibilities opened up by digital innovation. The analysis of fabrication technologies examines production processes in order to identify unit manufacturing processes within the succession of processes. Two projects are examined thoroughly. Frank Barkow and Regine Leibinger designed the gatehouse of the Trumpf factory near Stuttgart with a wide metal roof, protecting the open space to welcome visitors and employees. The space is not interrupted by the glass parallelepiped of the building, whose vertices are the pillars which support the 65 ft (20 m) of overhang of the roof. The transparency of the body of the building, which houses the reception and waiting room, is not immaterial, but rather emphasises the properties of the glass in the search of a semi-transparent effect. The way the structure is built exploits the possibilities of digital manufacturing and intends to be the visiting card, the demonstration of the expressive and constructive capacity of the machine tools produced by Trumpf itself as applied to creating the building which gives access to the factory. Manufacturing involved intensive mass change NC processes with cutting and punching, deformations with bending as well as the final assembly. Randall Stout won the international design competition for the expansion of the Art Gallery of Alberta, Canada. The exhibition spaces are highlighted by juxtapositions and intersections between two blocks: the grounding of the units of the building, accentuated in the cladding in local stone, and the complete transparency of the atrium, intersected by the sinuous form inspired by the aurora borealis, a characteristic of the latitudes of Alberta. This form has no centre, starting point or point of symmetry on which the eye can rest. The morphology, observed from different angles in crossing the piazza, the approach path or the atrium, appears changeable to the observer. The firm of Randall Stout Architects modelled in three dimensions the design of the complex geometry of the aurora borealis and of the building and the numerous intersections between the two using Rhinoceros (Rhino) software. From the

Rhino model, Empire Iron Works created the Building Information Modeling of the aurora borealis in Tekla Structures. Zahner's Design Assist Group imported and translated the geometric model into a Pro/ENGINEER parametric model. The parametric model of the secondary structure and of the cladding panels integrates the requirements of the manufacturing process and the assembly tolerances, specific to the ZEPPS™ system, and automatically generates the subsystems of the joints and of the panels necessary to make the complete roof.

From the start of the industrial revolution, when large quantities of metals of uniform quality became available, architecture has undergone intensive research and experimentation. New types of structures began to be developed, supported by new alloys and by innovative processing procedures designed to overcome increasingly higher resistances.

More than two centuries of research and development have produced a vast choice of metallic materials and manufacturing processes useful in giving form and structure to architecture. This chapter considers the state of manufacturing technologies in the light of possibilities created by digital innovation.

The previous chapter introduced the vast number of heterogeneous manufacturing processes made possible or enhanced by the integration of computers and software. The analysis of fabrication technologies examined the production processes, identifying unit manufacturing processes within the succession of processes.

This chapter turned then to the consideration of manufacturing processes used with metals, because

> The use of machinery in the cutting of metal into precise shapes involves, to begin with, a relatively small number of operations (and therefore machine types): turning, boring, drilling, milling, planing, grinding, polishing, etc. Moreover, all machines performing such operations confront a similar collection of technical problems, dealing with such matters as power transmission, control devices, feed mechanisms, friction reduction, and a broad array of problems connected with the properties of metals (such as ability to withstand stresses and heat resistance). It is because these processes and problems became common to the production of a wide range of disparate commodities that industries which were apparently unrelated from the point of view of the nature and uses of the final product became very closely related (technologically convergent) on a technological basis (Rosenberg 1976, pp. 156–157).[1]

Unit processes may be described in sufficiently general terms so as not to be constrained to working a specific type of metal or a specific component. Some processes, for example mass processing, use methods which also apply to other materials, such as stone or wood.

Metalworking processes are common in construction, more so than it might appear from an analysis of metal parts or components currently in use. They make load-bearing structures directly by means of frameworks, while they contribute to strengthening reinforced concrete, roofs and a wide range of equipment. Indirectly, on the other hand, metalworking processes are involved in the majority

---

[1] See Sect. 2.8.

of moulding processes and in the variety of uses of polymers in all building sectors, in weatherproofing, insulating, panelling and systems.

Working with dies requires advanced alloys and puts stringent requirements on the manufacturing processes, regarding tolerances of geometries, shape, position and finishing. They promote continuous innovation in new processes and tools and in the integration between CAD and CAM. Software is required to give a more accurate modelling and representation of the dynamic interactions between process, tool and geometry of the piece. Such interactions determine second by second the behaviour of the machine, the deviation of the tool and the stresses in the workpiece which together produce the "burring" in the part.

5.1 Mass-Change Processes

These are the most flexible and versatile manufacturing processes, due to their capacity to produce a variety of forms with different properties. Mass-change processes are usually defined as chip-making, since

> it is a separation process in which the shape or surface of a piece is changed by the use of a machine tool which carves away layers of material or chips. If there is no scrap or waste, the process is called net shape (Schulitz et al. 1999).

Casting also enables the making of complex shapes in large batches, but at the price of greater problems in achieving comparable accuracy and precision.

Processes for achieving mass-changes may be ascribed to two main categories, that is with the stock *rotating* or *fixed*.

5.1.1 Turning

Turning is, by definition, working with a stock which is rotated around its own axis. The piece worked on becomes conical or round, as in the case of nails, bolts or spherical joints.

Turning creates surfaces which are curved on the outside or the inside (for example threading), and also flat surfaces. The main movements of this process involve: the workpiece, which is continuously rotated around its own axis; and the tool, which is moved either by a thrust perpendicular to the direction of the cut, which determines the depth of the cut, or by a displacement parallel to the axis of the workpiece called feed.

The principal operations of turning are:

- *circular external*, in which the tool moves parallel to the axis of revolution;
- *facing*, in which the tool moves at right angles to the axis of revolution, permitting the creation of flat surfaces perpendicular to the axis of revolution;
  - *complex*, which involves the non-linear programming of tool movements parallel and perpendicular to the axis of the piece, allowing the creation of round shapes

Fig. 5.1 Horizontal lathe with working area 8 in 21/32 by 13 in 25/64 (220 by 340 mm)

according to mixed generators among straight lines in any direction, and lines on arcs of the circumference;

• *internal*, in workpieces with holes, it is possible to combine circular, facing or complex turning in the internal parts, achieving both curved and flat internal surfaces (Fig. 5.1).

5.1.2 *Drilling*

Drilling is related to turning, in that the rotating tool removes material from a workpiece which is held still, rather than moving. It is among the most common chip-making processes, as it is normally used in the manufacture of parts. The geometry of the tool, cylindrical, tapering and conical, corresponds to the shape of the process, which can also involve a combination of processes with different widths and faces, as in the case of drilling with conical or cylindrical countersinking.

The process of drilling may be broken down into the operations of centering, drilling, counterboring, boring and tapping. Because of the final geometry and the tolerance of the diameter and surface finishing, the process may require the combination of different operations with different tools. For example, centering has the aim of making a guide hole of small diameter used for positioning the tool, particularly when the drilling is on awkward surfaces or the axis is not perpendicular to the surface of the piece.

Fig. 5.2 Milling with a five-axis machine with tilting workshop furnished with multiple rotating piece-holders

The main movements of this process involve: the continuous rotating cutting of the tool; the drilling, on the axis of the tool; the placing and positioning of the stock or of the tool to bring the axis of the tool in line with that of the hole.

5.1.3   *Milling*

Milling is a process in which the tool moves relative to the piece, therefore stock or tool remain still. Movements of processing and feed, imposed on the tool relative to the piece or the other way round, may be linear or curved. The rotating tool, called the milling cutter, removes material from the piece in the form of a shaving. The form of the piece which may be achieved is multi-faceted, with different types of surfaces processed, from flat to ribbed with straight, helical and curved grooves, or with smoothed edges. Complex parts may be made, such as cogwheels, even though dedicated machines are more productive; or keyseats, which serve to join two pieces by means of self-blocking wedges, particularly used in the distribution of forces on an established perimeter of the workpiece.

The principal movements of milling involve cutting, continuous rotation of the tool; milling, straight or curved, of the workpiece relative to the tool; and positioning, which establishes the thickness of the allowance to be removed and is executed by moving the piece or the tool (Fig. 5.2).

5.1.4 Grinding

Grinding is the process of the removal of material, in the form of very tiny chips. It is generally used in the final phases of manufacturing, either for polishing or to make high precision pieces previously worked on with other processes, which left behind an allowance with a thickness of tenths or hundredths of a millimetre.

The tool used has no predefined geometric shape and is called a grindstone. Grindstones in the shape of discs, cylindrical cups, bevels or cones are used, depending on the processing needed.

The process is commonly used on pieces made with very hard or heat-treated alloys, for example following heat processes such as tempering or case-hardening.

The principal movements of this process are similar to milling; in grinding the tool is made to rotate at high speed (Fig. 5.3).

5.1.5 Machining Centres

Numerically controlled machining centres are multifunctional machines, which enable various processes to be carried out, for example milling, drilling, measuring, boring, counterboring and threading. A machining centre can therefore replace several single-function machines for chip-making processes.

The principal components of a machining centre are:

- *tool holder* mounted on the chuck to allow the automatic loading and unloading of tools; the tool-holder is the link between the chuck and each tool; it is necessary so that tools of different shape and size are interchangeable and can be automatically pre-lifted and mounted on the chuck;
- *tool magazine* used in order to have a wide variety of tools available for use with the machine; automatic switching between tools is performed using chain, drum, rack or alternating-arm magazines;
- *pallet* (moving table) for mounting the workpiece, useful for automatically transferring the workpiece into the working position, minimising the down time needed to change the workpiece; at least two pallets are used, one carries the workpiece and holds it in the working position, the other is transferred outside the tool's operating zone for the tasks of loading the piece to be worked and unloading the finished one;
- *working axes controlled*, usually between three and five; for example the figure shows the functioning of a machining centre with four axes; in the case being considered, the chuck can move along the Y axis, while the workpiece can move along the two linear axes X and Z and around the axis of rotation B.

The spread of machining centres with five or six axes is due to the growing complexity of workpieces with freeform surfaces. Unlike three-axis centres, they make it possible to complete processing on all sides of the piece with a single fixing to the pallet, to employ shorter tools and to make the tool always work at the best angle for cutting. On the other hand, machining centres with five axes require more complex programming, particularly in the phase of generating the route of the tool. CAM software programs have begun to implement new processing strategies (for example zig-zag, Z-plane) aimed at producing a more accurate surface finishing. Moreover, processes using five axes involve serious issues of tool collision with the workpiece. To avoid collisions, with consequent damage to the tool, the chuck or the workpiece, CAM systems have begun to implement simulation systems for routes and for obstacles to the tool and chuck (Fig. 5.4).

5.1.6   Erosion

Erosion differs from the processing methods considered up to now in this chapter, in that it does not use a moving cutting tool, rather it removes the material from the piece with the aid of mechanical, thermic, electrical, chemical or magnetic energy, or more often combinations of these (Jain 2008). Some processes carried out by erosion are counterboring, cutting and grinding.

With mechanical energy, material is removed by a high-speed jet of fluid, in which abrasive powders may be dissolved. The fluid is accelerated to high speed by means of pumping or ultrasonic vibrations.

With electrochemical energy, on the other hand, electrical energy is applied in combination with the chemical energy to enable the removal of material. Electrically

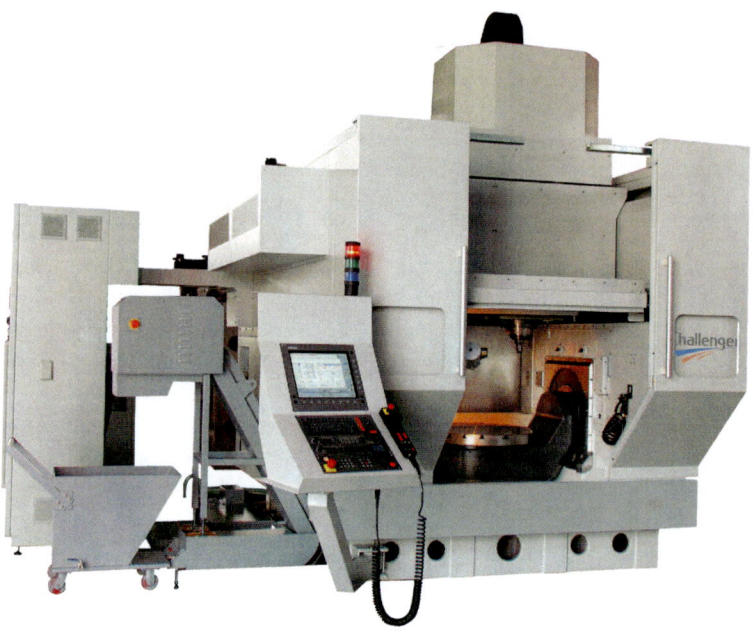

Fig. 5.4 Five-axis vertical machining centre with work area 26 in 3/8 by 32 in 9/32 by 23 in 5/8 (670 by 820 by 600 mm)

conductive materials are eroded by the reaction between the tool, which acts as a cathode, and the workpiece, which becomes the anode, by means of placing between them an electrolytic fluid under pressure.

With thermic energy, the material is removed as it reaches localised high temperatures which melt it or vaporise it. Because of the high temperatures, superficial phase-changes may be induced, which require further working processes. The principal heat sources used are electrical discharge, electron beam, plasma or laser.

Finally, with magnetic energy, magnetic abrasive powders are accelerated by the induction of magnetic fields and used as means of cutting or finishing.

The process of erosion may be taken to the micro or nano scale, for example to control surface finishing at a molecular level.

5.2   Phase-Change Processes

Casting is one of the most common techniques for shaping metals. The manufacture requires that the metallic material be heated until it reaches the liquid or semi-fluid state and then poured into a mould, in which it solidifies when the crystal grid of the material rearranges itself according to the new shape.

After the piece has been taken out of the mould, it may undergo further mass-change processes, often including chip-making.

An important part of casting is the design and making of the mould in which the liquid metal is to be poured. In order to shape the cavities which receive the casting, it is customary to use a *pattern*. Processes of preparing the mould (forming) distinguish between *investment moulds*, commonly used in earth, usable for only a single casting because the extraction of the workpiece destroys the mould, and *permanent moulds*, made so as to allow the extraction of workpieces without damaging the cavity, allowing them to be re-used for subsequent castings.

5.2.1   Investment Moulds

The preparation of the mould starts with the creation of the *pattern*, which is then inserted into sand or heat-resistant materials in order to form the cavity into which the molten metal will be poured. The shape and surfaces of the pattern are the same as those of the final piece.

The process of making the pattern depends on several inter-related factors: the geometry of the piece; the alloy used for the casting; the number of pieces to be produced; the tolerances; the dimensions and the mass of the workpiece.

Commonly used materials are:

- wax or polystyrene, which remain inside the mould and are therefore called lost-wax patterns, commonly made with rapid prototyping or mass-change processes;
- wood, relatively easy to work with mass-change processes;
- metal alloys, made with rapid prototyping processes for medium-small series, or with mass-change processes for larger series;
- plastic materials, mainly with rapid prototyping.

With regard to the shape of the piece, particular care in the design is required relating to:

- sharp-edged corners, because they do not easily resist the rapid cooling of the alloy when it flows into the mould, which would become damaged and absorbed into the casting;
- vertical surfaces, which could be damaged when the shape is taken out of the mould. This is why the introduction of small inclinations, called draft angles, is preferred;
- shrinkage of metals as they solidify. Therefore the mould must be over-sized according to the coefficient of linear contraction of the cooling casting applied to the local mass of the piece;
- the tendency to form bubbles of the specific alloy during the pouring, which determines the minimum depths and the angles of curvature (Fig. 5.5).

Fig. 5.5 Frame in birch plywood, made with NC milling machine, for investment cast of 3 MW wind turbine

5.2.2   Permanent Moulds

The mould is designed and made to be re-used in medium-large series of castings. Usually the mould is made of metallic alloys, of steel or special cast-iron so that it can be used to repeat several identical castings.

Because of the high number of pieces produced, permanent moulds are starting to be used in automated production processes, which assume control of the casting (which is now more commonly pressurised casting) and taking the casts out of the moulds, as well as cleaning the moulds so that they can be reused.

Pressurised casting, combined with rapid or accelerated cooling of the casts, allows the creation of a particularly fine crystalline structure with improved mechanical and structural characteristics.

5.3   Deformations

Digital technologies have made innovations in deformation processes, allowing the continuous and direct management of individual processes, each one complex by itself. Moreover, they have allowed the different processes to be integrated into a combined process, which is actively designed and managed. CAD and CAM systems support the design and simulation of the overall process,

> from which it is possible to obtain an estimate of the variation in shape of the product in the different stages of deformation; of the distributions of temperature within the workpiece and tools; and of the values of the tensions and deformations in the different areas, not to mention a complete evaluation of the forces at work (Santochi and Giusti 2000).

So, they manage and directly control the development of the manufacture through the succession of phases, being able to check any variations and evaluate actions useful in optimising the process whilst the work is going on.

5.3.1   Rolling

Rolling consists in deforming unworked metal, in the form of ingots for example, forcing them between two cylinders rotating in opposite ways which generate friction in the process. As stated above,[2] the process can be carried out hot or cold, with consequent changes to the mechanical characteristics of the piece.

The rotation of the cylinders is governed by digitally controlled motors, which gradually vary the number of turns. Generally, to make the designed form, the metal passes through successive pairs of cylinders, called the rolling train.

The speed of the process is controlled by the integrated management of the work, permitting the speeding up or slowing down of the individual motors of the cylinders or of the initial unwinding and the final winding. The aim is to keep the workpiece actively progressing, changing it by individual parameters as it goes along (Figs. 5.6, 5.7, and 5.8).

5.3.2   Forming

The piece is plastically deformed, in particular by forces that press the material. Digital control is used for the guide parameters of the manufacture, compression and temperature if appropriate, so as to achieve the best deformation of the metallic structure, refining large grain sizes into small grain sizes and causing the saturation of empty spaces and smoothing out inclusions with the result of improving the structural properties of the piece.

5.3.3   Forging

Raw material of an equivalent volume to the designed workpiece is plastically deformed so that it can be shaped into the desired form. Forging may be carried out cold or hot, but always at a temperature lower than the melting point of the

---

[2] See Chap. 3.

Fig. 5.6 Schnell Wire Systems cold-rolling lines for the production of smooth and ribbed wires for reinforced concrete

Fig. 5.7 Cassette for rolling or profiling operations, consisting of three rollers bent at 120°. The rollers cause the reduction of the section and, on request, of the ribs in relief

Fig. 5.8 Electronic control through Programmable Logic Controller. It manages the different parameters of working of the rolling machine: (1) set up or control of the wire feeding speed; the speed of the line depends on several factors e.g. the quality of the wire, lubricant, quality of the rollers, experience of the operators and mostly on the spooling unit; the motors are set to steady torque and power, each controlled by frequency converters; (2) set up of the final weight of the spooler according to the wire; (3) control of the pressure of the carriage device to prevent wire breakage while in tension; (4) electronic set up of the spacing between each layer of the spooler; (5) production detection; (6) maintenance detection

workpiece. Metals, before melting point, reach a malleable state far removed from breaking point, in which they can be easily deformed and worked and even undergo significant mass transfers.

Pieces obtained by forging have better mechanical properties than equivalent ones produced by the removal of shavings because the crystalline structure of the metal is deformed and displaced, but not broken. Therefore they can bear higher stresses. For example, forging is used in building for

> making wires in the form of metallic chains often forged for cables or for wire rope clamps (Schulitz et al. 1999).

CAD-CAM allows the best use of design and processing. By simulating manufacture it becomes possible to optimise the quantity of raw material needed for the shape of the piece, improving the filling out, avoiding production defects, like "folds", due to too much material, making an excellent size of grain in the metal, creating the best forging profiles for the allowance desired, reducing weight and costs. During the production process, forging cycles are digitally controlled, automatically managing forces and, if needed, heating at individual areas of the workpiece.

5.3.4   Extrusion

This is a plastic deformation process, carried out cold or hot, used especially in the production of semi-finished pieces including those of varying thicknesses, for example sections for windows and doors. The raw material, subjected to compression forces through a press, is forced to exit through a draw-plate, which moulds the section into the shape and size corresponding to the design. Tolerances and surface finishing are satisfactory so as not to need further processing. Usually it is done with alloys with high deformability, such as aluminium and copper.

The design of the draw-plate is particularly important for the outcome of the process. Normally it is assisted by CAD-CAM technologies, so as to make it in high-performance alloys.

Numerical control of the extrusion process automatically manages:

- the pressure of the press, so as to keep it either within the parameters of the project or steadily below the working threshold of the machine so as to reduce energy consumption and wear and tear on the motors;
- the temperature of the material;
- the speed of extrusion, determined by a combination of pressure and dragging speed;
- special functions at the start and finish of extrusions for reloading and emptying the press.

5.4   Joining and Consolidation

Joining processes have the purpose of assembling different parts into a single workpiece by means of links which are *reversible*, i.e. made with nails, bolts, rivets, pins and screws, or *permanent*, i.e. welding and brazing.

5.4.1   Welding

Welding is the technique used for joining two or more pieces. It is currently employed in the workshop and on the building site and can be carried out using different procedures chosen on the basis of the particular characteristics of the process: the type of alloy to be welded, thickness of the workpieces, position of the weld, type of joint and structural requirements of the joint.

In welding, a bond, called a joint, is created between the metals of the workpieces to be joined, if necessary with the aid of a further metal incorporated in the creation of the joint. A distinction is made between *autogenous welding*, in which the metals of the workpieces participate in the creation of the joint by melting, and *heterogeneous welding* (or *brazing*), in which the metals of the workpiece do not melt and the formation of the joint depends on the solder, which must be different from the metals of the workpieces because it must have a melting temperature below that of the parts to be welded.

In building, the principal autogenous welding processes are achieved by:

- melting: via the combustion of a gas, obtained by mixing acetylene with oxygen; with an electric arc with a consumable electrode—when the electrode comes into contact with the workpiece, it generates the passage of a very strong current with localised heating of up to 3,000 °C, which fuses the material of the electrode with that of the workpieces, creating the joint; with plasma—heating can reach the point of vaporisation of the metal until it generates a through hole in the workpiece which, as it solidifies, forms the joint; with a laser, which assures a high level of control and precision in the extent of the area and in the temperature, allowing a choice of whether to bring the metal to a mellow state, to melting point or to vaporisation;
- pressure: the parts are joined by means of pressure, whereby the passage of a very strong current, or a strong friction, produces an intense heat, bringing localised parts of the metals to the mellow state, making the welded joint; it is particularly used in joining thin parts without high structural requirements.

Numerical control of the welding processes is achieved mainly by using industrial robots because welding techniques require control of movement in order to reach the different points where the joint is to be made as well as coordination with the source of energy used (Loureiro and Bolmsjo 2006).

Industrial robots are highly flexible machines, as it is possible to program the movement of the mechanical arm, or manipulator, which often has some anthropomorphic characteristics. Robotic systems are currently used for loading in phase-change, mass-change and deformation processes (IFR Statistical Department 2011). Although the processes are different, the functions undertaken by the robots are comparable, such that they may be considered all together in the following paragraph.[3]

Robotised welding stations use the following welding techniques:

- *arc welding.* The robot is required to move and position the torch to correspond to the areas to be welded and effectively to integrate the welding generator; the aim is to optimise the functioning and the management of the process by switching on, switching off and if necessary resuming the welding so as to make the best joint;
- *laser.* Besides assuring a high level of control and precision of the welding process, it can operate without contact (this is called remote laser welding), even from distances of between 20 in and 40 in (50 and 100 cm), being able to reach

---

[3] See Sect. 5.5.

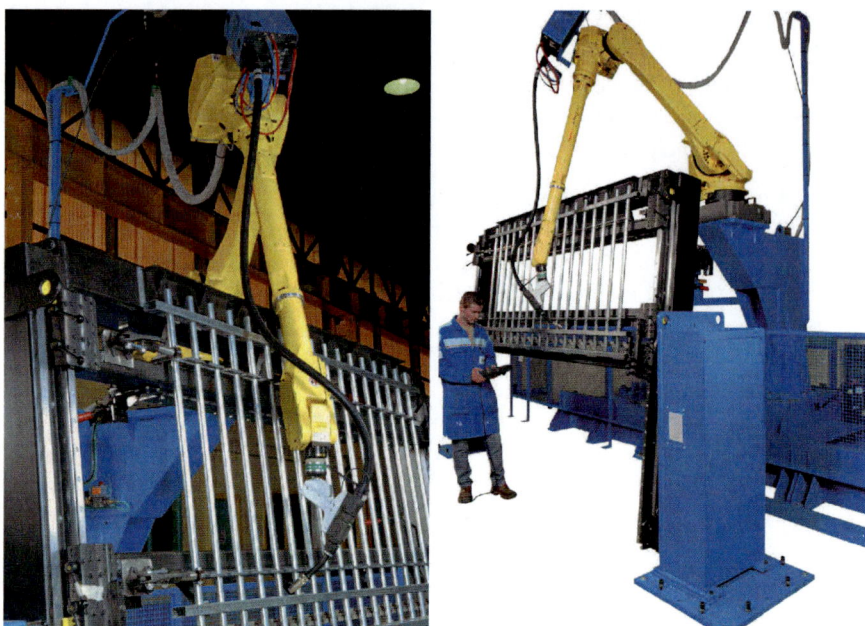

Figs. 5.9 and 5.10 Welding of metallic enclosure with six-axis anthropomorphic robot, configured for arc welding, mounted on rail of 315 in (8,000 mm) length

spots which are difficult to access and eliminating many arm movements, including corresponding down time.

Programming multi-axis robots for the welding process is approached either by learning (the operator makes the robot travel through the required trajectories and operations which the system memorises and repeats automatically), or by automatic generation from the CAD model of the parts. CAD-CAM integration assures better flexibility in the design and management of the process. The simulation must address the planning of complex trajectories across the field of work with the aim of positioning the welding tool in the best place with the least number of displacements, and that in a workspace encumbered by obstacles. Moreover, the simulation is required to quantify rapid accelerations and stops precisely, calculating the inertia of the manipulator and of the tools on it.

Metalwork processes which currently use different specific tasks (tracers, cutters and welders) and different machine tools can be automated even for small series with the integration of CAD-CAM and robotised workstations. A single robot can hold different tools, some for tracing, some for cutting and others for welding, programmed with a single process. Design software and integrated simulations manage the programming process with interruptions for the different tasks of the robot, allowing working time to be optimised and the best results obtained (Figs. 5.9 and 5.10).

5.5 Industrial Robots

The "principle of technological convergence" (Rosenberg 1976) is particularly appropriate to industrial robots, which carry out a vast number of manufacturing processes (phase-change processes, mass-change processes, deformations, joining, assembly and movement) by means of a relatively small number of homogeneous functions and operations.

ISO TR/8373-2.3 defines an industrial robot as:

> An automatically controlled, reprogrammable, multipurpose manipulator programmable in three or more axes, which may be either fixed in place or mobile for use in industrial automation applications.[4]

Industrial robots consist of a mechanical arm and a numerical control unit. The mechanical arm consists of a series of rigid segments connected together by flexible joints which have the function of positioning and orienting the mechanical arm in relation to its base. Types of flexible joints include: *prismatic* (telescopic), which allow linear movement between two segments without rotation between them; *rotary* (hinged), which allow revolving movement, in a single plane, of one segment compared with the other.

Robots can be catalogued according to the type and number of flexible joints they use.

5.5.1 Cartesian Robots

These can carry out movements in three planes created by prismatic joints. They are normally made with a structure fixed to a gantry, hence the alternative name *gantry robot*, or with a mounted structure in which a vertical or horizontal mount is anchored. Because of the presence of fixed elements, the structure is usually crush-proof, assuring reduced inertia, good energy performance, high precision and repeatability.

They are used where it is necessary to function across large work areas, perhaps moving large or heavy pieces, or along a sequence of several unit processes. They are also used for assembly, in which case they are limited to vertical and horizontal planes.

5.5.2 Cylindrical Robots

These can carry out movements in two planes and rotation around an axis, usually the vertical one. This rotating movement, which circumscribes a cylindrical working envelope, contributes to the name of this category of robot. Because of jolting

---

[4] ISO Standard 8373:1994, Manipulating Industrial Robots.

segments, in order to reach high precision a heavy and crushproof structure is necessary, which comes at the expense of speed and acceleration. They are used in loading machine tools, particularly those of medium-small size.

5.5.3 Anthropomorphic Robots

The movements are similar to those of a human arm, which inspires the name anthropomorphic as well as the naming of the different parts: foot, the bed; body, the main structural base; shoulder joint, the first joint; arm, the segment linked to the shoulder; elbow, the joint linked to the arm; wrist, final joint (generally endowed with several degrees of freedom), which can hold several tools interchangeably (Figs. 5.11 and 5.12).

The most common anthropomorphic robots have six degrees of freedom; this secures maximum versatility, allowing each point on the working area to be reached from any direction and with any orientation of the wrist. The arm works very jerkily, and is therefore exposed to significant bending moments. It is also subject to very variable moments of inertia, because of the load on the wrist as regards extension, acceleration and deceleration. These factors tend to limit the achievement of very high accuracy and repeatability, necessary for example in some processes involving the removal of shavings.

The high versatility of anthropomorphic robots makes them particularly suitable for replacing people in high-precision, repetitive or dangerous tasks such as:

- loading working units or centres, by loading and unloading workpieces from the pallets, and changing tools;
- arranging randomly placed workpieces, by means of a vision system;
- joining parts using reversible joints, made by screwing or riveting;
- joining parts using permanent joints, made by gluing, welding or brazing;
- assembly operations, including those with a high number of parts, including the movement of the mounted piece;
- mass-change processes, such as drilling, trimming, burring, slicing;
- mass-change processes based on erosion, laser cutting or water jet cutting;
- phase-change processes, by means of managing the casting, extracting the casts even at high temperatures, manipulating moulds, patterns and workpieces in an integrated fashion, so as to make working areas completely automated.

5.5.4 SCARA Robots

These can carry out movements of rotation around two axes and linear movement in one plane, usually the vertical one. This kinematic movement allows the arm to reach every point in a plane of work within the circumference of a circle. The movement around a column assures a crushproof structure with reduced inertia, high

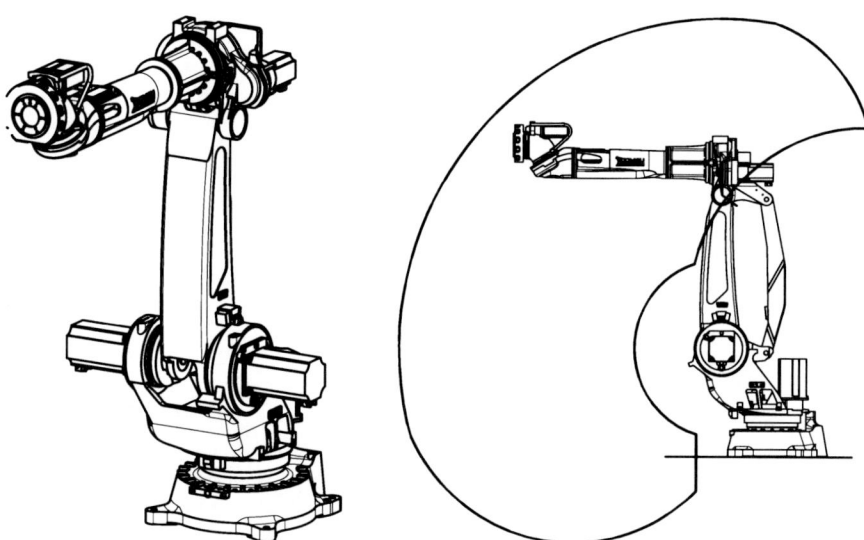

Figs. 5.11 and 5.12 Comau SMART-5 NJ4 anthropomorphic robot with six degrees of freedom and maximum payload at the wrist 170 kg. The wrist is hollow in order to contain all the wires. The kinematic structure is designed to reduce weight, inertia and interferences

precision and repeatability (in the order of hundredths of a millimetre). These robots are particularly suitable for the movement and precision assembly of small parts. Hence their name Selective Compliant Assembly Robot Arm (SCARA).

5.6   Barkow and Leibinger, Gatehouse in Stuttgart

In the project of the expansion of the Trumpf factory near Stuttgart, Frank Barkow and Regine Leibinger designed the reception building with a wide sheet of metal covering 35 by 12 yd (32 by 11 m): a wide, protected open space to welcome visitors and employees, not interrupted by the glass parallelepiped of the building, whose vertices are the pillars which support the 66 ft (20 m) of overhang of the roof (Figs. 5.13 and 5.14).

The transparency of the body of the building, which houses the reception and waiting room, is not immaterial, but rather emphasises the properties of the glass in the search of a semi-transparent effect. The glass walls are made with a cavity of 10 in (25 cm), filled with glass cylinders. The cylinders vary in diameter from 1 in 37/64 to 5 in 33/64 (40–140 mm) and are arranged in layers within the cavity, their diameters increasing from the base upwards: like grains which, shaken through a sieve, tend to arrange themselves according to their size; the smallest at the bottom, the largest at the top (Fig. 5.15).

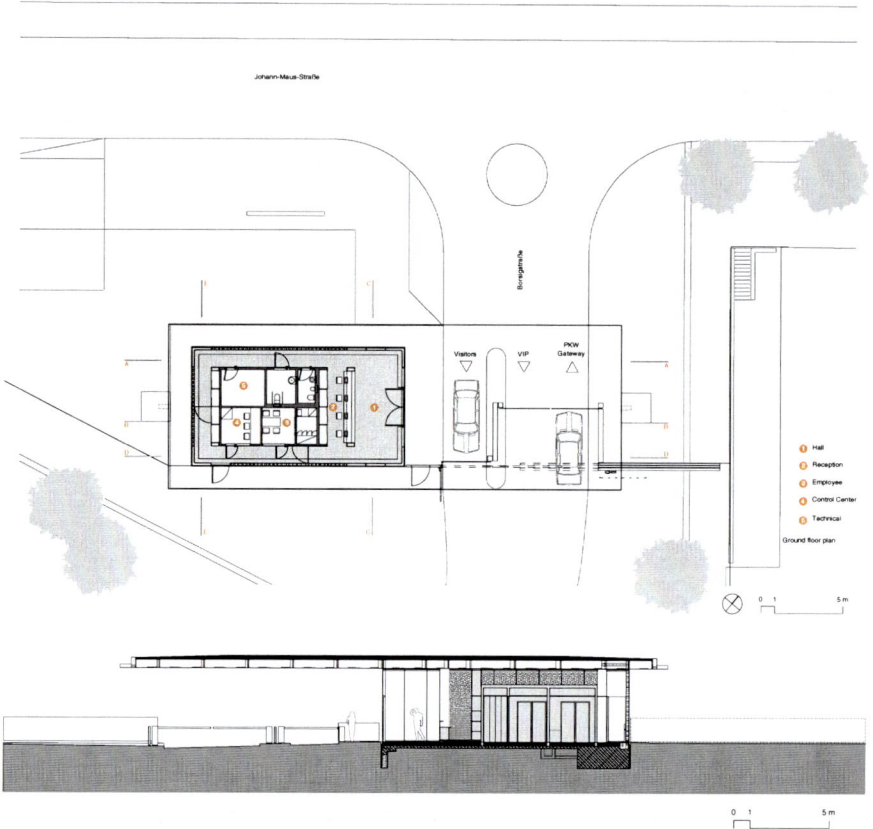

Figs. 5.13 and 5.14   Barkow Leibinger Architects. Plan of the reception building, the entrance to the Trumpf factories. Longitudinal section of the reception building

The light shining on the glass wall shows the diminishing density towards the top and produces

> a spectacular effect of the wall growing thinner. Its density gradually diminishes as it moves upwards and the semi-transparency effect moves downwards, so that the volume seems to dematerialise, lose consistency, mass and weight, emptying out. The "solid" is eaten away, reduced, thinned, and made lightweight. The operation is based on the subtraction of material (Marchesini 2008).

The prominent overhang of the roof, jutting out over the road to create a 66 ft (20 m) long porch, is subjected to a similar lightening process.

Making the structure lightweight, through a cellular steel framework, means that the roof is only 19 in 11/16 (50 cm) thick. In the design, the porch echoes the designs of Frank Lloyd Wright, who

> adopted the porch for his houses—not, however, encircling his buildings with it, but pushing it forward, in keeping with his cruciform or elongated plans, as an extension of the

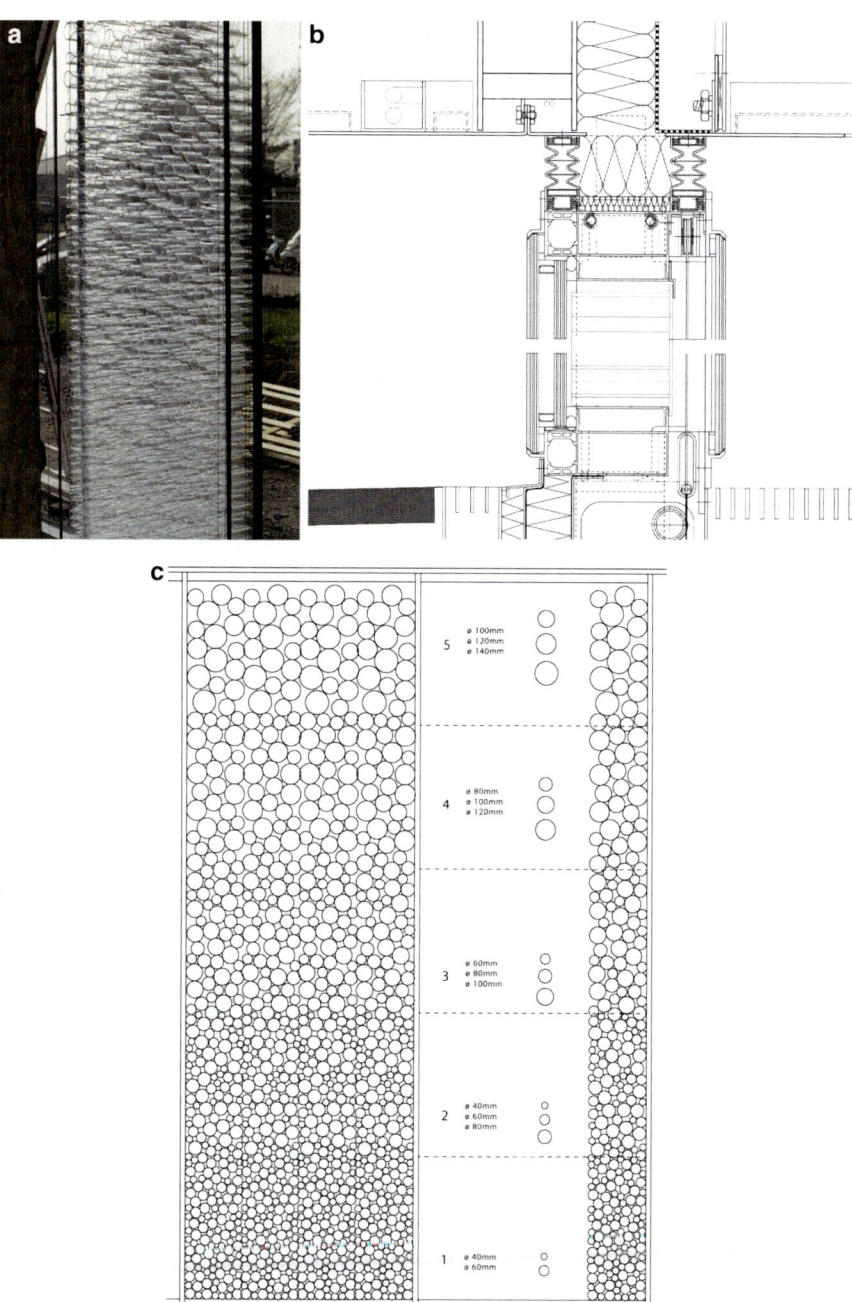

Fig. 5.15 (**a–c**) The glass walls are made with a cavity of 10 in (25 cm), filled with acrylic glass cylinders

wings. Very often it thrusts out into space as a pure cantilever hovering above the earth. Such a treatment had never been attempted before. True, it is the old element of the porch, but it is not simply something attached to the house; rather it is an essential part of the structure, molded as an inseparable part of it (Giedion 1967, pp. 480–409).

The roof forms an enormous honeycomb panel of more than 3,767 ft^2 (350 m^2) made of 1/8 in (3 mm) thick sheet steel. The connecting elements, in between the two soffits of the roof, increase the value of the bending rigidity of the roof, which is considerably higher than that that of a roof made with only two steel sides, though at the expense of a modest increase in material. This structural design shows similarities with that conceived and put into practice by the Centre des Nouvelles Industries et Technologies in Paris, in which the double shell is consolidated and reinforced by ribs in the intermediate layer of aeronautical design (De Nardi 2000).

The mass is lightened, however, by a perforation process which also has the aim of exhibiting the structural form and the load-bearing function of the underside of the roof. The lightening scheme expresses the stress diagram: the density varies with the moments in the plane, bigger at the centre, smaller towards the edges. Barkow and Leibinger undertook research into appropriate patterns to increase the intelligibility of the plan of the forces acting on the roof. The outcome makes evident the existence of tensions in the roof and gives shape and unity to the building (Figs. 5.16 and 5.17).

The way the structure is built exploits the possibilities of digital manufacturing and intends to be the company's visiting card, the demonstration of the expressive and constructive capacity of the machine tools produced by Trumpf itself as applied to creating the building which gives access to the factory.

The cellular structure of the covering is made of box girders, 28 in by 20 in (70 by 50 cm) wide and 10 ft (3 m) long (Figs. 5.18 and 5.19).

The perforation of each segment of box girder was designed according to the specific moments to which it must react. Barkow and Leibinger used RhinoScript and Grasshopper to model each of the subsystems involved in the design: the tension diagram, different geometric plans, the manufacturing processes, the assembly arrangements and the properties of the steel.

The restrictions and the specific possibilities of variation of each of the subsystems are implemented into the parameters and the algorithms of a generative model. The architects, interacting with these parameters, were able to generate dozens of variations.

Explorations within the broad domain of the various design solutions finally converged towards a design idea which was deemed satisfactory.

Manufacturing involved mass change processes with cutting and punching, deformations with bending and then the final assembly.

The steel sheet was cut using a machine tool equipped with a 5,000 W powered laser (TruLaser 3030).

Such power allows the rapid cutting of various alloys, for example construction steel and stainless steel, without introducing deformations and guarantees smooth edges so that the workpieces are ready for subsequent manufacturing processes. Numerical control automatically manages the power of the laser based on the cutting speed. The integrated control of the head displacement speed and the

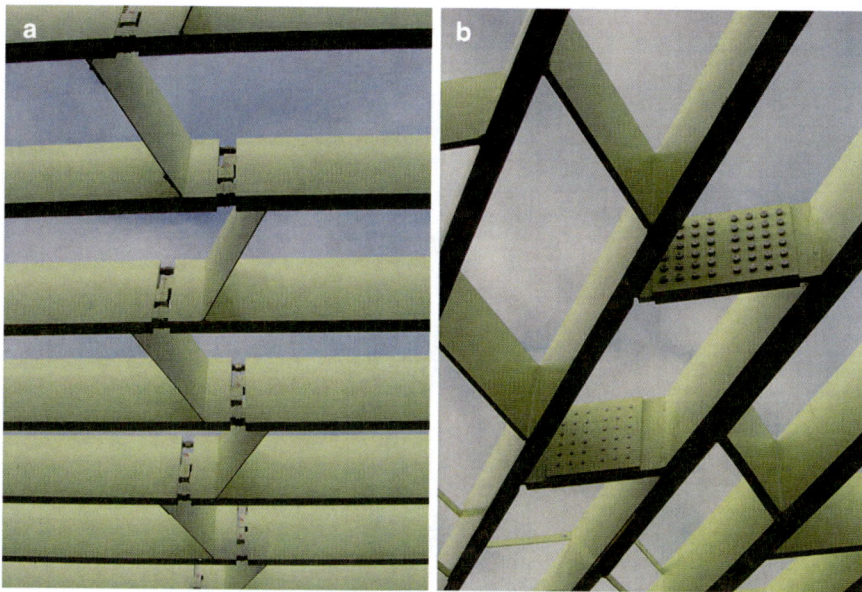

Fig. 5.16 (**a, b**) Girders assembled into a 24-in-thick (60 cm) honeycomb, load-bearing structure; the intermediate stiffeners contribute to stiffness of the 105-ft-long roof (32 m)

Fig. 5.17 Diagram of the tensions in the underside and topside of the roof

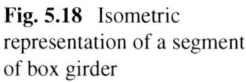

Fig. 5.18 Isometric
representation of a segment
of box girder

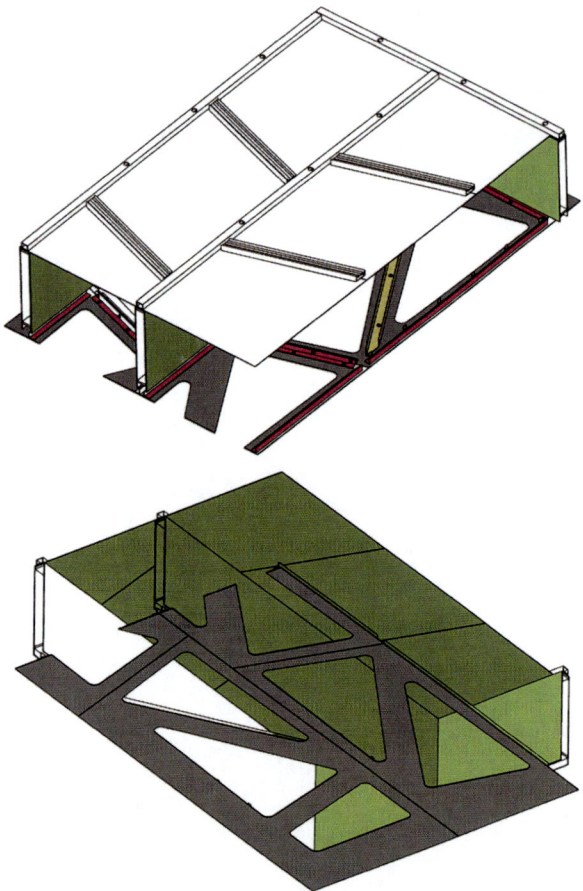

modulation of the power of the laser allows the best cutting parameters to be
obtained, even with small parts (Figs. 5.20, 5.21, and 5.22).

The software of the machine makes the preliminary checks, closes the open out-
lines, cancels out any duplicated cutting movements present in the CAD model, and
then proceeds to optimise the movements of the laser beam. Furthermore, it combines
the individual workpieces in a nest so as to make the best use of the sheet, both optimising the surfaces occupied by the parts and minimising the unused spaces between
the workpieces. Subsequently, the software defines the rounding of the corners and the
working cycles. Finally, it calculates the cutting trajectory and the processing
sequence: head displacements are optimised by keeping movements to a minimum.

The steel sheets, cut to the designed shape and size, are folded by means of
numerically controlled bending-presses (TruBend 3000). The exact bending angles
are made automatically by means of free bending, which gives the ability to fold the
sheet according to angles programmable at anything between 30° and 179°. The
different bending angles are made without having to change tools thanks to the

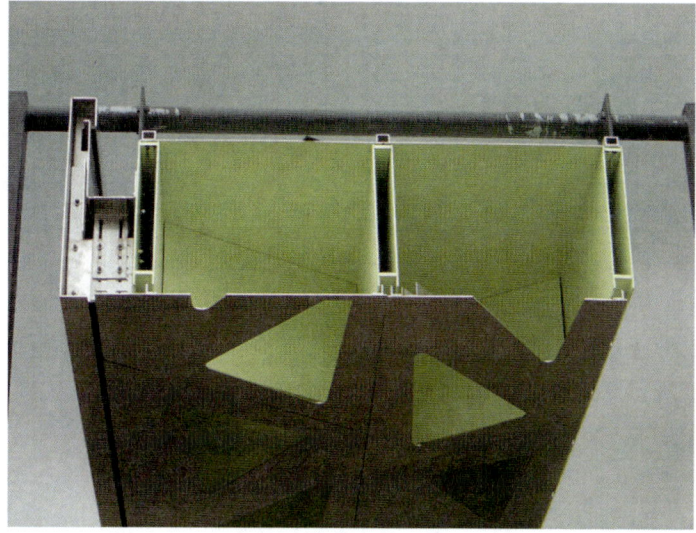

Fig. 5.19   Prototype of a segment of box girder

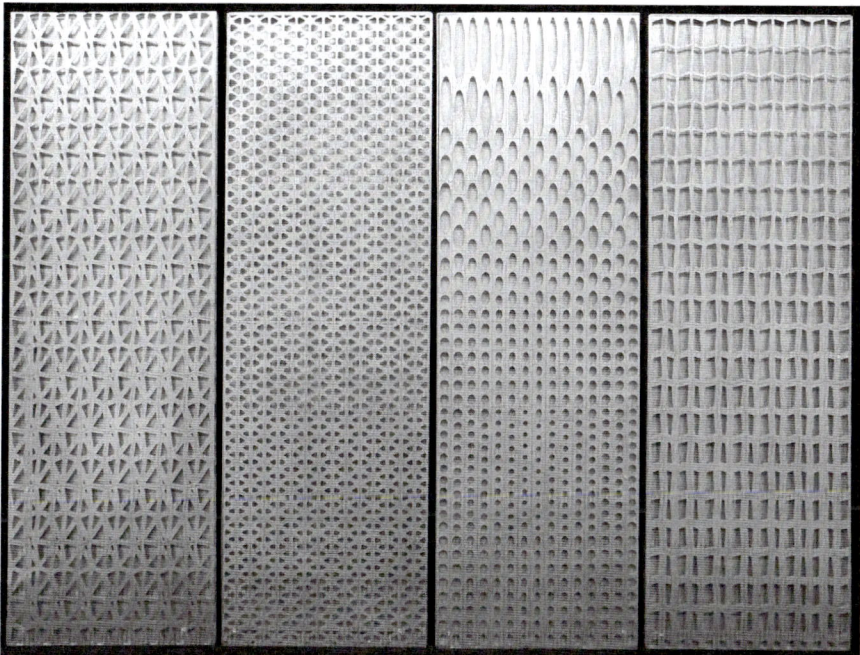

Fig. 5.20   Study of patterns representing the tensional forces of the roof

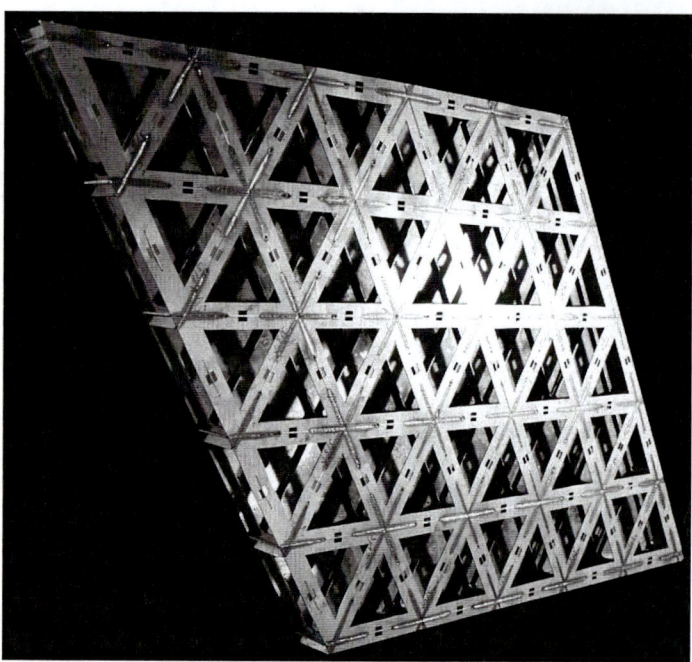

Fig. 5.21 Prototype of the cellular reinforcement of the roof

Fig. 5.22 The steel sheet was cut using a machining centre equipped with a laser (TruLaser 3030). The laser has 5,000 W power and allows cutting without introducing deformations and ensures smooth edges

Fig. 5.23 The steel sheets are folded into the final shapes by means of a numerically controlled bending-press (TruBend 3000). The exact bending angles were made automatically by means of free bending, which gives the ability to fold the sheet according to angles programmable at anything between 30° and 179°

digital system for measuring the angle patented by Trumpf in 1997, which enables checking and, if necessary, correcting the bending angle in real time (Fig. 5.23).

The software (TruTops) programmes every bend according to the type and angle of the bend, the thickness of the metal sheet and the properties of the materials and the tools as obtained from a specific database. According to these parameters, the system calculates the precise radii and shortening values for the manufacturing of each workpiece.

The software can simulate and display the complete bending process. Thus the machinist can check the sequence of processing units in advance (Figs. 5.24 and 5.25).

Punching prepares for the threads, which expedites the assembling of the individual parts into the cellular structure. All the various processes of threading, forming, holing, and deburring are punched with a single stroke. The appropriate tool is loaded in the head of the machine (TruPunch 3000) and is rotated around the axis to adapt it to the specific process: for instance, to make rectangular or round holes or to cut circles and curves.

The presser foot holds the sheet in position and prevents it from undesired deforming during the process. The exact control of the tools allows the elimination of imperfections during the punching: the precise execution prevents any "burring", i.e. the formation of uneven edges in the cut, yielding a smooth edge. At the end of the assembly process, each individual girder was unique, designed and made to be

Fig. 5.24 All the various processes of threading, forming, holing and deburring are punched with a single stroke by the machine (TruPunch 3000). The threads are predisposed with the punching, which expedites assembly of the individual parts into the cellular structure

Fig. 5.25 Assembly in the shop floor of the segments to construct the lengthways girders of the roof

Fig. 5.26 (**a, b**) At the building site, the box girders were positioned and bolted together to assemble the complete honeycomb load-bearing structure, which was raised and hinged to the four columns

positioned at a specific place in the structure. The complete cellular structure of the roof was assembled from girder segments 10 ft long (3 m), 28 in wide and 20 in high (70 by 50 cm). In the shop floor the segments were bolted together lengthwise to form a single girder the length of the roof, 105 ft (32 m).

Transported to the building site, the boxed girders were positioned and bolted together to make the complete cellular roof, which was then raised and hinged to the four columns (Fig. 5.26).

5.7   Randall Stout Architects, Art Gallery of Alberta

The international design competition for the expansion of the Art Gallery of Alberta, Canada was won by the firm Randall Stout Architects, Inc. in 2005.

The winning project planned to renew the existing block of the building, designed in 1969 by the architect Don Bittorf, raising it by two floors and grafting onto it a wide glass atrium extending towards the centre of the city.

The exhibition spaces are highlighted by juxtapositions and intersections between blocks. The grounding of the units of the building is accentuated in the cladding on the bottom at the sidewalk in local stone, which intends to display the lithic materiality with the mass and solidity of blocks, like excavations in a quarry (Figs. 5.27, 5.28, 5.29, 5.30, 5.31, and 5.32).

The complete transparency of the atrium is intersected by the sinuous form inspired by the aurora borealis, a phenomenon specific to the latitudes of Alberta (Fig. 5.33).

Randall Stout has captured the changeable and evanescent nature of the aurora borealis which, in the lithe harmony of the inorganic forms, permeates and enfolds the atrium, continuously wrapping itself inside, outside and through the large glazed area.

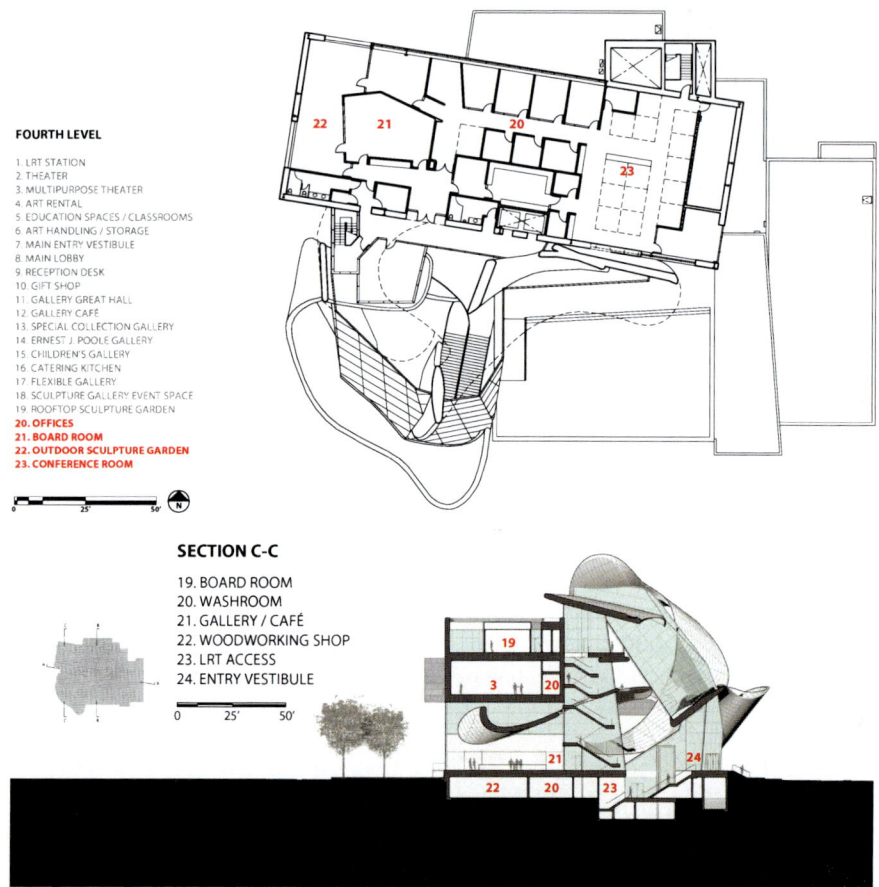

FOURTH LEVEL

1. LRT STATION
2. THEATER
3. MULTIPURPOSE THEATER
4. ART RENTAL
5. EDUCATION SPACES / CLASSROOMS
6. ART HANDLING / STORAGE
7. MAIN ENTRY VESTIBULE
8. MAIN LOBBY
9. RECEPTION DESK
10. GIFT SHOP
11. GALLERY GREAT HALL
12. GALLERY CAFÉ
13. SPECIAL COLLECTION GALLERY
14. ERNEST J. POOLE GALLERY
15. CHILDREN'S GALLERY
16. CATERING KITCHEN
17. FLEXIBLE GALLERY
18. SCULPTURE GALLERY EVENT SPACE
19. ROOFTOP SCULPTURE GARDEN
20. OFFICES
21. BOARD ROOM
22. OUTDOOR SCULPTURE GARDEN
23. CONFERENCE ROOM

SECTION C-C

19. BOARD ROOM
20. WASHROOM
21. GALLERY / CAFÉ
22. WOODWORKING SHOP
23. LRT ACCESS
24. ENTRY VESTIBULE

Figs. 5.27 and 5.28 Randall Stout Architects. Plan of the fourth floor of the museum. Section of the atrium and the exhibition spaces

The concept of movement has its origins in the search for artistic innovation at the beginning of the nineteen hundreds in the experimentation of Bruno Taut, Vladimir Tatlin, Herman Finsterlin and Le Corbusier who

> searched in organic forms like the spiral for an example for the study of continuously growing structures (Fanelli and Gargiani 1998).
>
> As emblematic evidence for this, I suggest the gigantic double spiral of metal lattice designed in 1919 by Vladimir Tatlin, a painter, designer and architect among the founders of the Constructivist movement in Russia. In his model made for the Third International, we can recognise the tendency of all new architecture to dissolve the volume and to create a spatial penetration between outside and inside (Vittoria 1994).

The form has no centre, starting point or point of symmetry on which the eye can rest. The morphology, observed from different angles in crossing the piazza, the approach path or the atrium, appears changeable to the observer.

Figs. 5.29, 5.30 and 5.31   The new regional Museum of Alberta rises up in the centre of the city of Edmonton, Canada

Fig. 5.32   Cladding in the severe local stone accentuates the massiveness of the exhibition building, emphasised by the recessed joints

Fig. 5.33 Aurora borealis

The tenuous glow of the aurora borealis is reflected by means of the careful placing of complementary materials of moderate contrasts, like zinc, stainless steel and
aluminium (Fig. 5.34).

Zinc, with the patina of age, has soft reflections and nuances which vary from
greenish, blue, and greys, from cool greys to warm greys. It has a low level of reflectivity, so as to interact delicately with the tenuous changes of light and colour of the
northern latitudes.

The outside of the aurora borealis is made of stainless steel. The steel surface
is etched with micro-engravings of varied angling and depth, using a proprietary
process from Andrew Zahner Company, sold with the trademark Angel Hair™.

Distribution, angling and depth of the finishing are calculated according to the
desired level of reflectivity, optimised to eliminate glare and made using mass-
change processes and numerically controlled water jet engraving on the surface of
the steel (Figs. 5.35 and 5.36).

The underside of the aurora is made of white-painted aluminium.

During the day, the surface of the aurora borealis shields the light bearing down
on the hall. Here the stainless steel creates tenuous reflections which change
incessantly with variations in the sun's height and position. During the long night
of these northern latitudes, the glazed area tends to dissolve and to highlight the
nocturnal reflections on the external surface of the stainless steel. Meanwhile the
internal surface, in white aluminium, reflects the illumination of the hall, creating
a diffuse and soft low light and contributing to mellowing the direct lighting
(Fig. 5.37).

Fig. 5.34 The wide area of the atrium crossed and interwoven with the surface of the aurora borealis

Fig. 5.35 Angel Hair™ panel detail of the engravings on the surface of the stainless steel made with numeric control water jet

Fig. 5.36 Assembly on site of the Angel Hair™ cladding

Fig. 5.37 The external surface of the aurora borealis in stainless steel highlights the nocturnal reflections, the internal surface in white-painted aluminium reflects the direct illumination of the hall, creating a soft low light

Fig. 5.38 The contact sections between the hollow tubes are calculated exactly so as to control the cutting precisely. The aim is to obtain the best area of contact between the tubes for sake of the welding

The surface of the aurora borealis totals 2,045 ft^2 (190 m^2), curving within and outside the area of the atrium and around the south-west corner of the museum. The aurora borealis is supported by an internal structure of hollow steel tubes with a diameter of up to 16 in (406 mm) and a thickness of 1/2 in (13 mm). The processing methods of the structure may be classified as mass-change, deformation and joining processes.

The tubes, made by rotary piercing seamlessly so as to increase their structural properties, first pass through a machine tool which cuts them to exact lengths. Then a numerically controlled tube-bending machine curves them according to the defined radius. The bend die, clamp die and pressure die contribute to the bending process: the bend die forms the given radius of bend and prevents the tube from flattening; the clamp die holds the tube in position during the process; the pressure die presses the tube into the bend die, while the pressure should be controlled continuously in order to manage the correct force against the bend die for the duration of bending. The ends of the tubes are cut according to the shape of the contact section, to be joined subsequently by means of arc welding. In order to join tubes with different bending radii, it is important that the shape of the contact section be calculated exactly, so as to precisely control the cut.

Only under these conditions is it possible to obtain the best area of contact between the tubes. In this way, the welding process does not cause tensions or deformations when girders and tubes are joined to compose the different sub-framing structures into which the whole aurora borealis is subdivided (Figs. 5.38, 5.39, and 5.40).

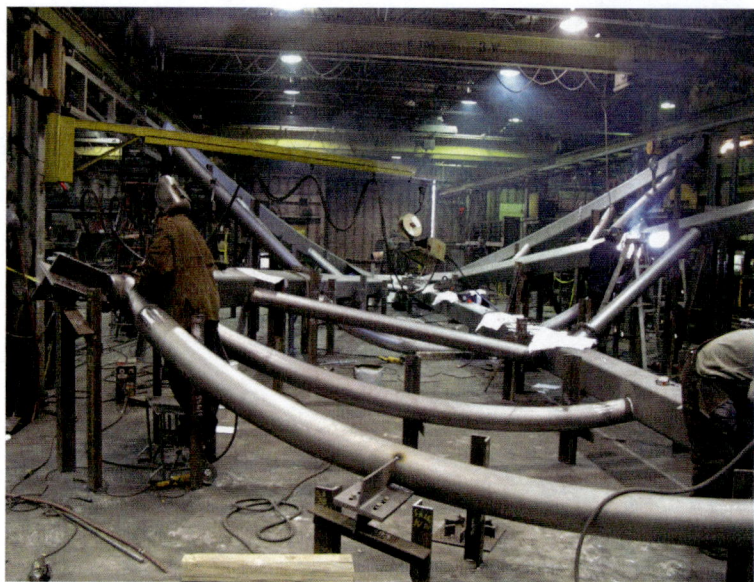

Fig. 5.39 Empire Iron Works proceeds to the arc welding of girders and hollow tubes in order to put together the different subframing structures into which the complete aurora borealis is divided

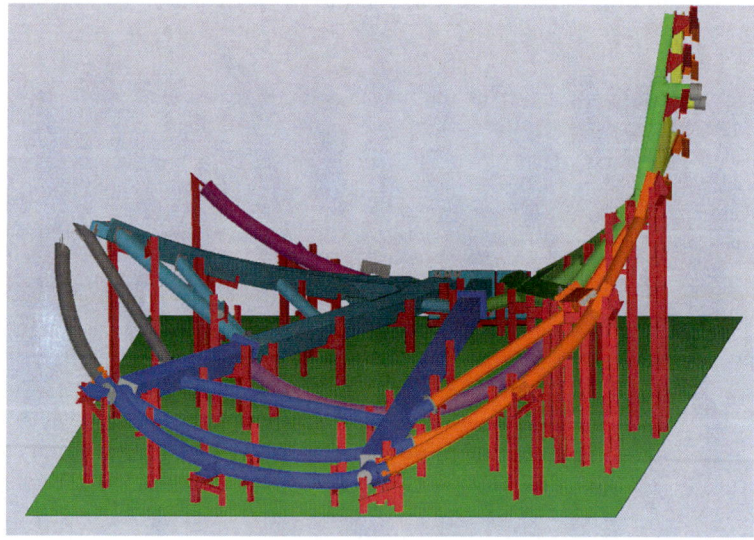

Fig. 5.40 Model in Tekla of the sub-framing structure

On the building site, the sub-framing structures are joined together and to the building using flanges and clamping bolts.

The aurora borealis crosses the glass cover at various points, creating critical nodes where structural continuity must be guaranteed and thermal breaks made. With external temperatures which commonly reach 40 °C below zero, it is essential that the thermal bridge be interrupted, because apart from heat loss, they can induce a high level of internal condensation with a consequent rapid deterioration of the steel. The problem of these nodes was resolved by inserting into the structure, next to the glass wall, a 1 in (25 mm) thick block of wood. The block was tied to the flange by means of bolts of structural steel of small diameter and high-strength, ASTM A325, which is sufficient to limit heat conduction (Figs. 5.41, 5.42, 5.43, and 5.44).

The firm of Randall Stout Architects modelled in three dimensions the design of the complex geometry of the aurora borealis and of the building and the numerous intersections between the two using Rhinoceros (Rhino) software (Fig. 5.45, 5.46, and 5.47).

From the Rhino model, Empire Iron Works created the Building Information Modeling (BIM) of the aurora borealis in Tekla Structures version 13. The Tekla model was used to check the correct alignment and orientation of the girders and of the hollow tubes in the structure. To this primary structure was added the secondary structure for fixing the cladding panels. The secondary structure, the support clips and the panels were made with Zahner Engineered Profiled Panel Systems (ZEPPS™).

The BIM in Tekla allowed checking and dealing with interferences between the primary and secondary structures and the cladding. Moreover, it was used to establish the angles necessary for the alignment of the support clips for the panels, particularly in critical areas such as the edges of the aurora. Tekla was also used to design the clamping masks which can be rotated on their own axis so as to return the joints between the girders to right angles. This simplified the manufacture of the girders and their assembly in the workshop (Figs. 5.48 and 5.49).

From the Tekla models of the single parts of the primary structure, the engineers of Empire Iron Works generated the numerical control programs needed for the manufacturing process, for the laser cutting of the girders and of the hollow tubes and for the bending of the hollow tubes. The coordinates of the vertices of the single principal elements of the substructure were extracted from the model in Tekla and loaded into the memory of the total station Trimble TS415. The software of the total station, after pinpointing the actual position of the top of a girder or tube, compares it with the designed position and calculates any alignments and oddities, which is useful for precisely fixing the element into the structure. The repeatability of the measurements is of 5 in degrees (1.5 milligons), appropriate to the precision of numerical control process. After soldering, the coordinates of the individual elements of the structure are surveyed. The measurements, imported from the total station in the BIM in Tekla, were used for the management and progress checking of the assembly process in the shopfloor (Fig. 5.50, 5.51, and 5.52).

Figs. 5.41 and 5.42 On the building site, the sub-framing structures are joined together and to the building by means of flanges and clamping bolts

The Tekla model of the structure assembled in the workshop was exported by Empire Iron Works in IGES and imported by Zahner into its own CAD/CAM system Pro/ENGINEER™. This model served Zahner as a reference point for designing and manufacturing the joining elements in the secondary structure and the cladding panels.

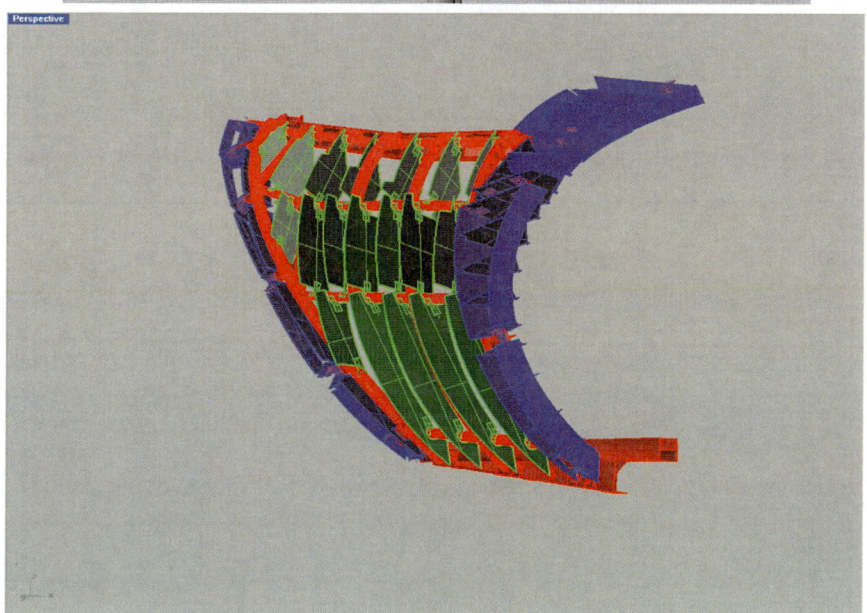

Figs. 5.43 und 5.44 Model in Tekla of the structure and substructure

For shop drawing reviews the models in Tekla and Pro/ENGINEER were exported into IGES to import them back in Rhino.

Zahner's Design Assist Group translated the geometric model, imported into Pro/ENGINEER, into a parametric model of the secondary structure and of the

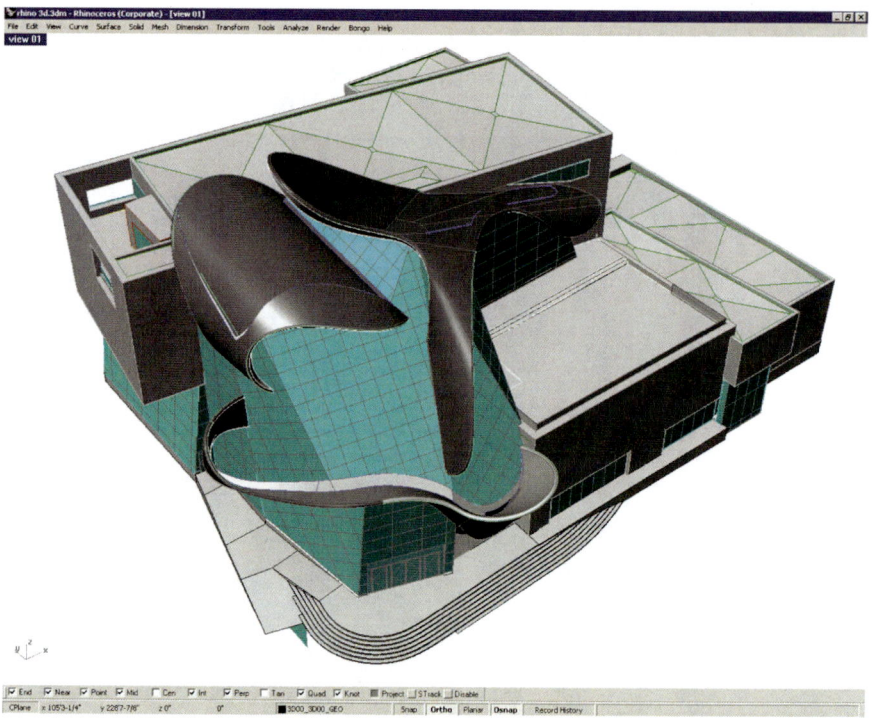

Fig. 5.45 The model of the skin in Rhino

cladding panels. The parametric model integrates the requirements of the manufacturing process and assembly tolerances, which are specific to the ZEPPS™ system. The parametric procedures developed by Zahner automatically generate the subsystems of the joints and of the panels necessary to make the complete roof. In the generation, the system can consider further parameters, such as the joints between the panels, the method of connection with the principal and secondary structures and the presence of flashings.

The parametric model is put under simulations of the static and dynamic loads acting on the roof as produced by pressure from wind, the weight of snow or earth tremors.

The parametric model also integrates the requirements of the manufacturing processes, allowing the direct generation of programs for the CNC machines, eliminating the need to introduce changes, adjustments or reprogramming of the machining centres in the workshop.

Zahner manages and manufactures each design according to the "Just in time"[5] philosophy, reducing the time taken for design, manufacture and construction. The parts and components of each project, including that of the Art Gallery of Alberta,

---

[5] See Sect. 3.1.

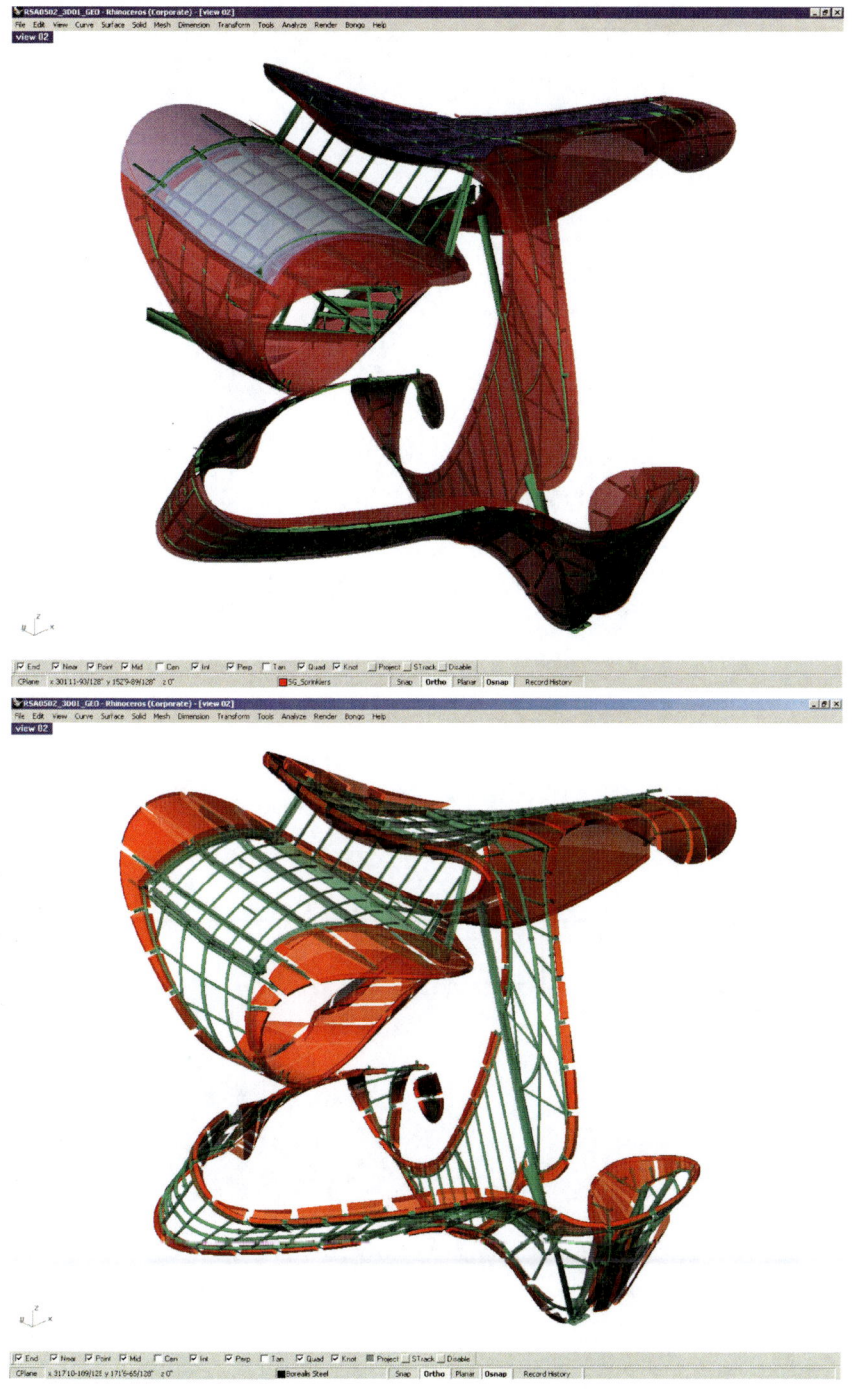

Figs. 5.46 and 5.47   The model of the structure and the cladding of the aurora borealis in Rhino

Fig. 5.48 The model of the structure and of the substructure in Tekla

Fig. 5.49 The structure during construction

Fig. 5.50 The total station leads the precise fitting together of the elements of the structure during assembly at the shop floor. After the welding, the coordinates of the individual elements of the structure are surveyed in order to build the CAD model of the assembled structure

are made to fit the requirements of the specific project. In this way, warehouse storage is eliminated in favour of the direct processing of the lots for the current job.

Zahner assiduously updates the parametric procedures so as to formalise tested experiences with design procedures, production and construction that are acquired with new projects. The parametric system becomes the firm's consolidated knowledge base. In this way, Zahner improves automation in support of the whole process, from design to construction. The aim is to optimise not only individual stages, but also the integration between the phases of the overall process: the parametric design allows the interactive adaptation to the flexible production in real time with regard to both assembly and construction.

The ZEPPS™ system shows several structural and manufacturing similarities with aeronautical constructions: it has a secondary structure in light materials like

Figs. 5.51 and 5.52   The subsystems of links and panels of the aurora borealis, the joints between the principal and secondary structures

aluminium, with preformed panels fixed to the substructure. Overall, the ZEPPS™ system is optimised and integrated according to the principles of "lean manufacturing,"[6] in which the design of the individual parts and components is aimed at real-time production requirements of pre-assembly in the workshop and of assembly on the building site (Figs. 5.53 and 5.54).

---

[6] Ibid.

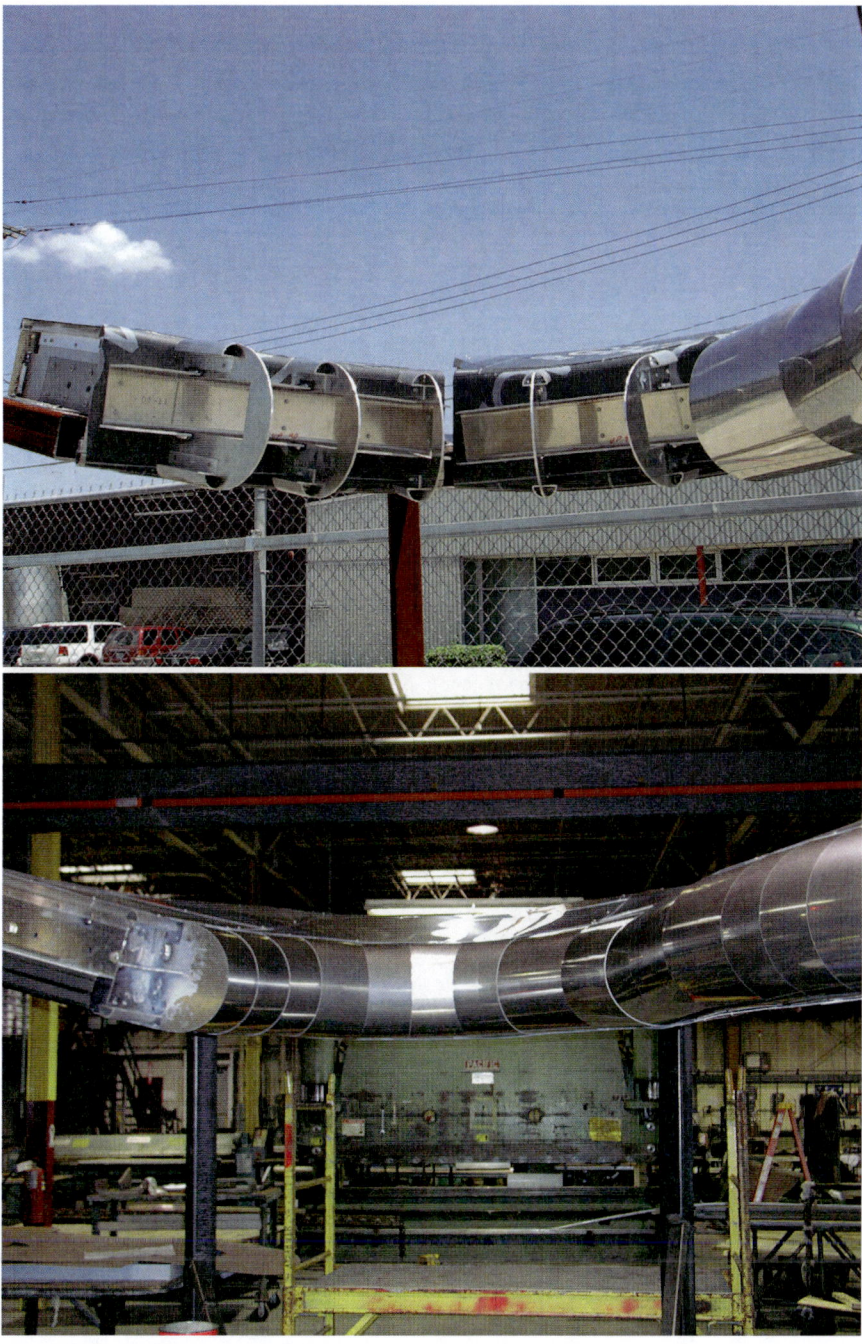

Figs. 5.53 and 5.54 Edge of the aurora borealis, details of the secondary structure and of the joints of the panels in the ZEPPS™ system

Zahner's manufacturing plant covers 64,600 ft^2 (6,000 m^2) and contains a numerical control machine tool for water jet abrasive cutting with a usable working area of 10 by 20 ft (3 by 6 m). The machine can process parts with a wide variety of dimensions (from the single stud up to a girder) and materials (it can cut or saw aluminium, copper, stainless steel and special steels with thicknesses including anything from a few tenths of a millimetre to 150 mm).

References

De Nardi D (2000) Jean Prouvé: idee costruttive. Testo & Immagine, Torino

Fanelli G, Gargiani R (1998) Storia dell'architettura contemporanea. Spazio, struttura, involucro. Laterza, Bari

Giedion S (1967) Space, time and architecture the growth of a new tradition. Harvard University Press, Cambridge

IFR Statistical Department (2011) EUnited robotics, "world robotics 2011 – industrial robots". VDMA Verlag, Frankfurt

Jain VK (2008) Advanced (non-traditional) machining processes. In: Davim JP (ed) Machining: fundamentals and recent advances. Springer, London

Loureiro A, Bolmsjo G (2006) Welding robots: technology, systems issues and applications. Springer, London

Marchesini M (2008) Barkow Leibinger architect. Gatehouse Materia 58:132–137

Rosenberg N (1976) Perspectives on technology. Cambridge University Press, Cambridge/New York

Santochi M, Giusti F (2000) Tecnologia meccanica e studi di fabbricazione. Cea, Milano

Schulitz HC, Sobek W, Habermann KJ (1999) Atlante dell'acciaio. Utet, Torino

Vittoria E (1994) Il costruttivismo progettante. In: La Creta R, Truppi C (eds) L'architetto tra tecnologia e progetto. FrancoAngeli, Milano

Chapter 6
Digital Woodworking

Abstract This chapter considers the evolution of the wood construction industry through methods of integrated automated fabrication, thereby achieving higher environmental sustainability, in particular in building systems which have been able to transform local materials into components with high technical performance. Five key unit manufacturing processes are identified and examined: (1) and (2) mass-change with and without the removal of wood shavings respectively, (3) machining centres, (4) deformations, (5) joining and consolidation. Two projects are thoroughly examined. The Jürgen Mayer H. firm designed the winning project, the Metropol Parasol, for the International Competition on the re-design of the Plaza de la Encarnación in Seville. The project developed around interrelated layers of public spaces. The idea of the design is to create a huge sunshade—literally a parasol, hence the name—, a forest moulded in an organic shape, which should create in a natural way an urban micro-climate favourable to the activities taking place in its shade. The parasol is cut from composite timber panels with layers of veneer placed lengthwise and crosswise to the length of the panel. A three-dimensional digital model of the entire structure of the parasol integrates all the panel widths and the dimensions of the metal connectors. This model was imported from Finnforest Merk, the timber contractor, into the CAD/CAM software system bocad-3D. From the 3D model of the structure, the system automatically worked out the NC program for controlling the Kuka robot with seven axes. For the work of each panel, the anthropomorphic robot carried out the unified manufacturing processes: cutting, milling and drilling. Reconstruction of Bari's Teatro Petruzzelli took place in 2008 after its destruction due to an arson attack. The external dome and the roof were rebuilt as they were before the fire, preserving unaltered the original geometry and structural plan. A description is given of the rebuilding in glulam of the inner dome, which was originally built in timber and wattle-and-daub. Turning to a glue laminated timber system allowed the designers to use thinner widths, which are more efficient from the point of view of stability and consequently help increase the load bearing capacity for the equivalent span. The inner dome was pre-fabricated by making technological units and elements in the factory and later assembling them in a highly organised building site. One specific constraint and challenge for the management of the construction process was handling the parts to and into the building site. The timber contractor, Stratex, used a proprietary CAD/CAM system, developed in CadWorks, to automate the generation of pieces of the structure. The system generated the numerical control for the various unit processes of the production that were necessary to produce the finished pieces in a condition ready for assembly. Each piece of the structure, after the lamination processes of the beams or timber elements, proceeded automatically through the various processes of moulding, rubbing down, pre-drilling, matching and finishing.

Wood was the most important material for all pre-industrial economies. Nathan Rosenberg considers its substitution with fossil fuel in industrial activities to be

> one of the central events of the Industrial Revolution [...]. The increasing scarcity of wood and the desirability of substituting coal became increasingly clear in Great Britain as early as the second half of the sixteenth century, during which time the price of firewood rose far more rapidly than prices generally (Rosenberg 1976, p. 272).

In the United States during the nineteenth century wood was the principal construction material for dwellings and for the production of vehicles, in addition to being the main source of fuel.

> Woodworking was particularly vital for the United States, where the per capita consumption of timber in 1820 was five times higher than that of Great Britain (Thomson 2009, p. 51).

For these reasons the mechanisation of work processes developed relatively early: in 1835 machines had already been invented for sawing, turning, planing, tenoning and mortising. However,

> Iron and steel were substituted for wood across a whole range of investment goods in the nineteenth century, going back to the pre-Civil War period. Machinery, ships and bridges, made of wood in 1800, were made of iron or steel in 1900 (Rosenberg 1976, pp. 254–255).

In Italy, the construction sector began to adopt wooden structures on a large scale starting in 1920 due to the shortage of iron and steel, the importing of which was embargoed. Necessity stimulated the invention and adoption of composite beams made with small pieces of wood, which were therefore cheaper and light, designed to create buildings of large size (Tampone 2011, p. 3). In the construction sector, wood processing industries responded to the rise of materials such as concrete and steel with the development of innovative construction systems, a development supported by the automation of production processes.

> Further technological changes also generated methods of economising on wood needs, without replacing it with competitive materials, or with products of lower cost compared with more expensive woods, for example plywood and veneers. (In addition, wood waste is now utilized in the manufacture of fiber board and synthetics.) Further technological changes, such as for example the self-powered chain saw, the tractor and the truck have reduced the cost of extracting and transporting the timber from its forest stands. Finally,

other technological changes have, in effect, significantly increased the size of our forest resource base by making possible the utilization of low-grade materials which previously had gone unused. Until the 1920s the woodpulp industry utilized only the spruce and fir trees of the northern part of the country. Improvements in sulphate pulping technology during the 1920s made possible the exploitation of faster growing southern pine which was previously unusable, as a result of which the South accounted for over half of the country's woodpulping capacity by the mid-1950s (Rosenberg 1976, p. 255).

The development of the wood construction industry continues with mature industrial products by means of methods of integrated automated fabrication, the system of design and certification of efficiencies, higher performances and the necessary specialist skills to make these systems work. For example, take the panel, frame or window component,

> From a handcrafted product which for centuries had only been described according to the raw material used to make it [...] it was transformed into a building component, characterised by a use of mass, endowed with known and verifiable performance statistics, accompanied by a reliable set of technical information, to the point that today every window whether it be made of wood, steel, aluminium, PVC or any combination of these materials, is to all intents and purposes an industrial product, and not just because it is now produced industrially, but because it is conceived, designed, produced, described, distributed and certified as such (Sinopoli and Tatano 2002, p. 248).

Increased environmental awareness favours sustainable resources and wood is a favoured resource, in particular in building systems which have been able to transform local materials into components of high technical performance:

> The development of the sector then veered toward developing and using wood resources produced in the context of a short supply chain, with a specific interest in strengthening and diversifying the use of locally produced components (Callegari and Zanuttini 2010, p. 13).

6.1   Mass-Change Processes

In woodworking, mass-change process techniques have a large number of applications. They are used in the production of a multiplicity of forms with different characteristics. They are used in the activities of primary processing which turn the uncut round wood of the felled tree trunks into semi-finished wood products in the form of sawn timber, panels, veneer sheets, veneer slices etc. They are used in the second processing which turns the semi-finished materials into finished items and components. They are also widely adopted in process procedures which, with a single series of fabrication processes, turn the uncut wood directly into finished items, for example beams, planking, laths and posts.[1]

---

[1] Valuable technical documentation was provided by Antonio Cossio, Woodengineering team, and by Matteo Simonetta, Associazione costruttori italiani macchine e accessori per la lavorazione del legno [Italian association of makers of woodworking machines and accessories].

Fig. 6.1 Independent clamp log band saw and log carriage, with electronically controlled hydraulic blade push and blade tensioning unit, hydrostatic feed with speed of 98 ft 5 in 7/64 per minute (30 m/min)

6.1.1   Mass-Change Processes with the Removal of Wood Shavings

One of the main processes including the removal of wood shavings is sawing, in which the material is subdivided into items usually with flat parallel faces. The tool used is the saw: a blade with teeth which, thanks to movement relative to the wood, cuts the original block, removing a fragment, called a shaving.

The main types of saw are: chainsaw, bandsaw, reciprocating saw and circular saw.

A chainsaw works with a blade made of a roller chain with teeth inserted into its links. The chain is rotated by a motor, and it is particularly suitable for being moved about to fell trees and to remove branches from the trunk.

The band saw is made of a thin band of steel with teeth on the cutting edge. The band is a continuous blade, so that it can be moved ceaselessly between two rotating flywheels.

Often bandsaws are situated at the beginning of the production line for uncut wood in order to work on logs including those of large diameter and to prepare them for subsequent processes (Fig. 6.1).

Bandsaws are also used to "re-saw", that is to obtain thinner planks from a large plank.

The main advantage is flexibility in working. This flexibility is increased by electronic control, which permits the programming of different patterns of

Fig. 6.2 (**a**, **b**) Log debarker-ring type, for logs of between 6 in 19/64 and 31 in 1/2 (16 and 80 cm) in diameter. The cutting tools and the scraping tools carry out a separate pressure to adapt to the log

cutting, adaptable to peculiarities in the trunks, the types of planks produced, and the direction of tapering of the logs. Moreover, it makes a thinner cut with less wastage of shavings. Although computer numerical control has taken on a large part of the expert management needed for its operation, management and maintenance (particularly of the blade), it remains a complex and expensive machine.

Reciprocating saws show similarities with bandsaws in the flow of the blade, which is obtained by an alternating movement instead of a continuous rotating movement. They offer less flexibility compared with bandsaws, requiring constraints on the material to be worked on and in the assortment of shapes and sizes the saw can produce.

Circular saws are composed of a toothed disc which rotates rapidly to remove the shaving and effect the cut. It is probably the machine with the simplest operation and production. The working processes are constrained by the radius of the disc, which cannot be increased without introducing vibrations that debase the cut. To counter this, it is necessary to increase the width of the blade, increasing wastage in proportion to this.

Debarking machines are often used in the preparation of tree trunks. To produce uncut round wood which fits the needs of successive workings, the debarkers must change the pressure of the tools to suit the characteristics of the wood and of the size and shape of the log. The operation of the machines is increasingly managed electronically with sensors, processors and actuators (Fig. 6.2).

The milling of wood does not differ, methodologically, from that of other materials, notably that of metals[2]: the tool is in relative movement with regard to the piece with straight or curved movements of the cutter or the feeder. Substantially changing the properties of the material being worked involves changes in the tool's parameters of feed, rotation or both, as well as the shape and size of the cutters. It is possible to have a large variety of cutter profiles, for example moulded cutters to produce moulding, grooves or joints. Programming the parameters of movement, rotation and cutter type allows the working of rebates, frames, incisions, mouldings, grooves, as well as complementary shapes, tenons and mortises, which are used to unite two items with a joint.

6.1.2 *Machining Centres*

Machining centres spread rapidly in the activities of primary and secondary processing because of their flexibility and their capacity to use a single machine to carry out different processes: widening, threading, milling, drilling and measuring. To better respond to specific processing procedures which often coincide with the product specialisation of woodworking firms, machining centres were developed which were dedicated to the manufacturing of trusses, prefabricated buildings, doors and windows, beams and laminated beams.

One significant example is work stations for doors and windows adopted to optimise the specific recurring processes in the production of frames. These centres can achieve in a single working cycle all the main processing operations of windows and doors (end milling, tenons, sides, cutting at 45°, holes) thanks to the combined action between unit of milling and plane of work of the pieces. Planes of work are managed automatically by robotised systems which organise, place and fix the pieces according to the most efficient combination of dimensions, shape and processing. Moreover, they can be loaded and unloaded without down time, while the milling unit is busy doing something else.

The centres are also designed to carry out complementary processes in a single working cycle, for example holes for screws, usually entrusted to dedicated machines in successive phases.

Work stations for windows and doors ensure greater product flexibility, being able immediately to vary production between different types (shutters; doors; small external doors; stairs; windows), between series and special production as well as ever varying shapes and sizes, for example trapezoid, triangular, arched, with three centred arch, elliptical arch or round arch. Efficiency and flexibility promoted their rapid adoption both in small and medium enterprises, where they execute the entire production, and also in medium-large businesses for the manufacture of special casings and for the prototyping of new products before they are mass produced (Fig. 6.3).

---

[2] "Unit processes may be described in sufficiently general terms as not to be restricted to working with a specific type or a specific component. Some, for example mass-change processing, may also be methodologically significant for other materials, such as stone and wood". See Sect. 6.1.

Fig. 6.3 (**a**, **b**) Numerically controlled work centre for the production of windows and doors

Fig. 6.4 Rotary slicer for making sheets of up to 158 in (400 cm) long. Includes a thermoregulation group for the tools in order to avoid stains on the veneer surface and for remote control of the cutting parameters

6.1.3 Mass-Change Processes Without the Removal of Wood Shavings

To obtain thin sheets, usually a few millimetres thick, slicing or peeling machines are used after debarking the log (Giordano et al. 1999).

In slicing, a cutting tool is arranged along a plane so as to cut off a continuous cross section from the log: the sheet. After cutting one section, the tool or the log moves exactly the amount necessary for cutting the thickness of the next sheet. Usually the sheet thickness obtained by slicing is between 1/32 in and 1/16 in (0.5 and 1.5 mm), less than that achieved by peeling (Fig. 6.4).

Since in many cases the wood is hard, it is helpful to increase the plasticity of the material before working on it with appropriate heat preparation processes.[3]

In peeling, the logs are worked on radially. It is necessary that the tapering is first eliminated from the logs (rounding), and that they are then heat treated[4] in a similar way to being prepared for slicing.

To obtain the sheets, the cutting tool is kept in constant and continuous contact with the log, which rotates on its own axis. It is essential that the operation of the machine is continuously managed to avoid discontinuities. Today this is performed by electronic control systems which, from the laser scan of the log, program the best route of the tool and manage the continuity of the process without any interruptions. The high speed cutting of the sheet is monitored in real-time by the scanner, which transmits the status and quality of the process to the control computer. The computer constantly modifies the parameters of cut and rotation so as to maintain the system in the best specific operating conditions for the format of the programmed sheet.

Peelers can produce sheets of thickness varying from 1/32 to 13/32 in (0.5–10 mm), commonly between 3/64 and 7/64 in (1 and 2.5 mm). The sheets are currently used as layers in the production of laminated veneer lumber and plywood.

6.2 Deformations

Like other materials, wood expands if heated. While the values of the coefficients of expansion are lower than those of materials commonly used in building, in the presence of humidity wood becomes subject to greater expansions, particularly across the grain. The loosening of cohesion between the fibres, under the combined action of heat and moistures, gives the wood plasticity. In industrial processes, plasticity is obtained via immersion in hot water for long periods of up to several days so that it completely soaks into the parts to be worked, or with steam in order to reduce the time taken.

In slicing and peeling, the wood is first subjected to this process in order to increase its plasticity.

The process of inducing plasticity in wood was exploited by the company of Thonet Brothers of Vienna from 1829 in order to make items of bentwood furniture, among them the chairs which bear their name (Massobrio and Portoghesi 1990). Once the wood becomes pliable, items are bent into the desired shape using stainless steel moulds. After being cooled, the items are again heated, making the new shape permanent.

The commercial success of Thonet products was such that in 1869, when their patent on bending wood with steam ran out, several other companies specialising

---

[3] See Sect. 6.2.

[4] Ibid.

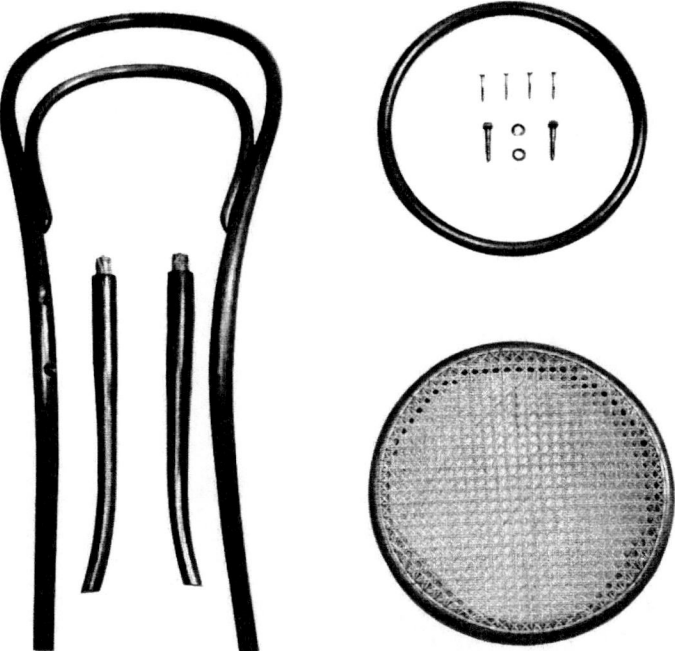

Fig. 6.5 Bentwood Thonet chair dismantled for shipping. This Thonet model is considered one of the most successful pieces of furniture

in the deformation and working of wood sprang up in that same city of Vienna (Sala and Sala 2005, Fig. 6.5).

The process of bending also found applications in building, for example in the creation of staircases. Yet wood thus treated has the limitation of not being able to be exposed to the elements, in that humidity rapidly changes the expansion and resistance of the material.

Hardwood items are bent with moulds subjected to forces of compression using a press. Control of the press is digital to keep it automatically within the design requirements and within the working threshold of the machine (Fig. 6.6).

In the last few decades ammonia-based treatments have been introduced for plasticising wood (Hoadley 2000). Ammonia is more efficient than steam because it interacts with the lignin and the cellulose component of the cell wall.

> Two basic systems have been developed: immersion in liquid anhydrous ammonia at atmospheric pressure, and treatment with gaseous anhydrous ammonia in closed chambers at 145-psi pressure. […] Upon removal from the liquid or gas treatment, the pieces can be bent to shape with little springback. Within minutes to an hour, the ammonia dissipates from the wood sufficiently to set the bend in place, and eventually stiffness and rigidity are restored virtually to original levels (Hoadley 2000, p. 179).

This process is faster and may also be applied to types of wood not treatable with steam, although at the expense of greater environmental impact.

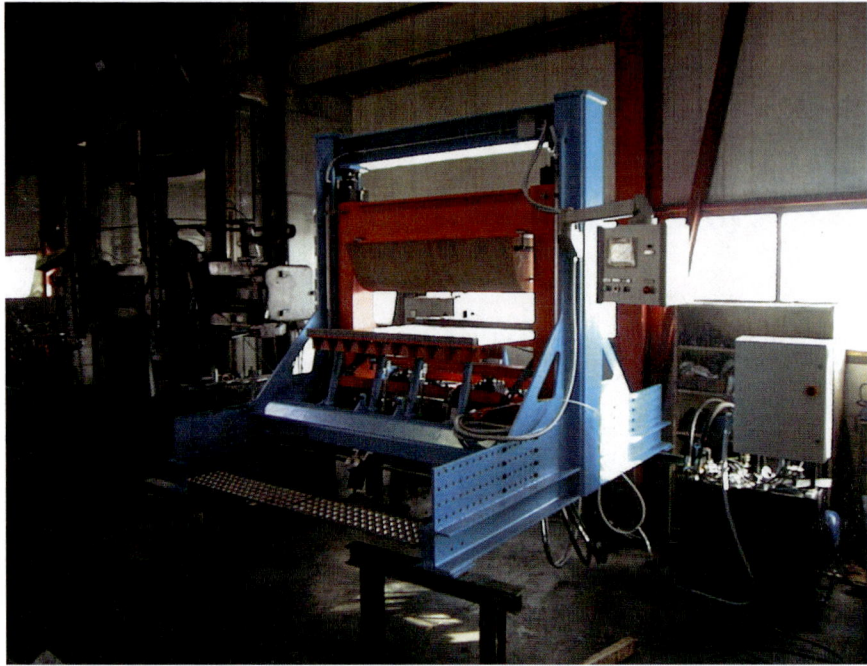

Fig. 6.6 Stretching press for bending items in solid wood of length varying between 2 and 118 in (50–3,000 mm) with maximum thickness of 3 in 1/2 (90 mm)

6.3   Joining and Consolidating

Processes of joining and consolidating play a central role in the manufacture of a wide range of products based on composite wood and reconstituted wood. Beginning with the innovation in binders, which made available a large variety of adhesives with different ways of laying, mechanical and chemical properties, the industry developed automated systems for the production of semi-finished or finished items and components.

From the end of the nineteenth century, the industry began to make products in reconstituted wood in order to improve performance compared with the natural material and to make use of the waste material from production and working which totalled about half the volume of the log. Today, virtually the entire log is used in production, doubling the volume of products achieved from the same mass of natural wood worked. This is made possible by the addition of binders and additives which work together to create products with better thermal, structural and mechanical performance (Lehmann 2011).

> We find a vast array of products based on reconstituted wood according to various methods: structures in wood veneer; boards of chipboard or MDF; multi-layered flooring materials; veneer faces for windows and doors. In the latter case the innovation was made

possible, as observed above, by the presence of increasingly reliable glues or, in more highly developed cases, of glues reinforced with fibres which give the wood all the properties of a solid block and also have very good fire retardant qualities. [...] The coupling and gluing of boards, laths or veneer sheets with increasingly reliable glues has allowed spans to be achieved which were formerly impossible when using wood in its natural state. In this field beams with a wide span are now mature products, whilst we begin to see innovative works in complex structures in which laminated timber no longer acts as a beam, but acts as a membrane or like a three dimensional shell to achieve domes and more curved surfaces (Sinopoli and Tatano 2002, p. 173).

6.3.1   Plywood

Plywood was chronologically one of the first composite timber products. A huge variety for many different purposes and performances is available on the market. Simple plywood is made by stacking sheets, called veneers, on top of each other, with the direction of the grains of adjacent veneers across each other. The veneers are glued together to form a sheet. Plywood panels are made in varying numbers of layers, from three to eleven or more.

The production cycle of plywood takes place in highly efficient automated production lines in which the uncut logs go in and the finished prepared panels come out. The cycle of production progresses from preparing the logs, to peeling and drying the sheet, to joining the sheets, and finally to gluing and consolidating with a press. Sometimes this sequence may be followed by the finishing and cladding of the panels.

Plywood is more stable both in shape and in size than solid wood, even though as it ages it may be subject to small splits in the outer layers due to the peeling process (Figs. 6.7, 6.8, 6.9, and 6.10).

6.3.2   Laminated Veneer

Laminated veneer timber is designed and produced so as to satisfy the most stringent structural requirements. It is also one of the most efficient ways of using wood in building structures.

It is not a recent product; the first patents on laminated veneer were registered in Switzerland and in Germany towards the end of the nineteenth century. A German patent of 1906 signals the real beginning of construction in laminated veneer timber. It is a technique subject to constant innovation; one significant step forward came with the introduction of waterproof binders based on phenol-resorcinol in 1942.

The production cycle of laminated veneer timber starts with the preparation of the boards in order to ensure that they remain the same size and the finishing, necessary for the joining of the veneers. It is essential to control the humidity of the material and to select boards with uniform mechanical qualities. Humidity is stabilised

Fig. 6.7 Peeling lathe for logs of large diameter with electronic control system for the machine and for the sheet

Fig. 6.8 Drying plant for green sheet with automatic loading and unloading

in dryers. The sorting of the boards is today carried out using a scanner which, after peeling on four sides, monitors each item, identifying any deformations or defects. Useful sections are salvaged from the rejected boards, with parts which do not meet the requirements being cut off.

During the selection process, sets are chosen of boards with similar mechanical behaviour and of equal width and depth, but of different length. The boards

Fig. 6.9 Gluing line for sheets with automatic loading and unloading

Fig. 6.10 Pressing machine
for consolidating panels
with automatic loading
and unloading

Fig. 6.11 Pressing machine for consolidating laminated beams with a maximum length of 105 ft (32 m), height between 63 and 86 in 1/2 (160 and 220 cm). The pressure exerted along the beam is controlled and managed electronically, also in order to adapt it to beams of varying thickness

are laid end to end to make up the desired length of the beam through joints, usually cross joints, finger joints or scarf joints, achieved by milling in continuous jointing lines.

After being passed through a glue applicator, the layers are automatically stacked together to make the required size of the beam, which is consolidated in a gluing press.

As they succeed in putting boards together end-to-end, side-by-side or face-to-face, the plants can produce beams of any dimensions. Limits derive only from the capacity of the presses and from the road transport of the finished product (Fig. 6.11).

6.3.3   X-Lam Panels

The X-Lam[5] system uses layers of locally sourced timber stacked at right angles and glued together. The cross-layered structure gives the panel excellent mechanical properties and high dimensional stability. These properties are intrinsic to the composite structure, and thus allow the use of structural wood classed from medium to low quality.

---

[5] Abbreviation for Cross Laminated Timber

Fig. 6.12 Semi-automated plant for the production of laminated panels of solid timber in cross-grained layers (X-Lam)

The production of an X-Lam panel starts with sawing the boards, shaving them on both sides, any necessary lengthwise jointing through finger-jointing, and joining the different layers together with polyurethane adhesives. Then, the panels are sawn into the shapes of the design, and if necessary also moulded. Openings are cut for the insertion of windows and doors. After this, holes for the tenon joints may be milled on the edges and factory-drilled holes for implants milled in the flat surfaces. Finally, as an additional cladding, plasterboard or plaster-fibre-board panels may be fixed to the fronts.

The complete process is carried out in partly or totally automated plants according to the needs posed by the volumes of production. The performance of the board is linked to the quality of the joining and consolidation process.

Particular care is required in checking the type of glue and in the calculating the load of the joints which must be done on representative samples and according to methods which conform to the class of service expected (Cremonini and Zanuttini 2010, p. 43).

In Germany the process of joining and consolidation using adhesives may only be carried out by certified companies.[6] For this reason, among small producers wooden plugs or metallic pins may be used alongside or in place of glue (Fig. 6.12).

---

[6]Certificate A, B or C DIN 1052:2008-12.

6.4   Jürgen Mayer H. Architects, Metropol Parasol in Seville

The re-design of the Plaza de la Encarnación in Seville began with the International Competition in 2004. The competition requested ideas for a new role for the plaza, which had formerly been destined to be an underground parking lot until excavations discovered an archaeological site of Roman origins. This led to the need to redefine the urban role of a location which used to be the central market in the historic heart of Seville.

The firm Jürgen Mayer H. designed the winning project, the Metropol Parasol. The idea developed around interrelated layers of public spaces. At the subterranean level a new museum is opening for the Roman remains; on street level is the space for shops and the market; thus the plaza is raised to 16 ft 4 55/64 in (5 m) above the ground. The public space extends beyond this in height as far as the balcony plaza at 70 ft 6 15/32 in (21.5 m), where bars and restaurants are opening and where there is the start of the overhead "architectural walkway" on the magnificent spatial structure (Figs. 6.13, 6.14, 6.15, 6.16, and 6.17).

Jürgen Mayer H. conceives a forest moulded in an organic shape, a fusion between parts, the trees, which lose their identity as individual items and become a single harmonious form placed in view of the architectural aim coordinating it. The structural intention is to set up a dialogue with the memory of the city, proposing a structure which aims to relate to the scale of the city's own identity. Jürgen Mayer H.

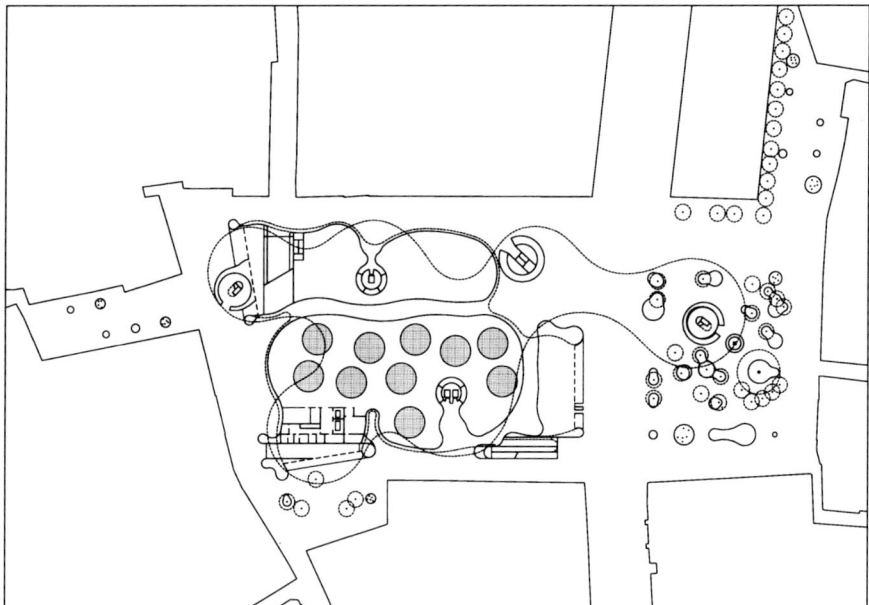

Fig. 6.13 The street-level with the shops and the market

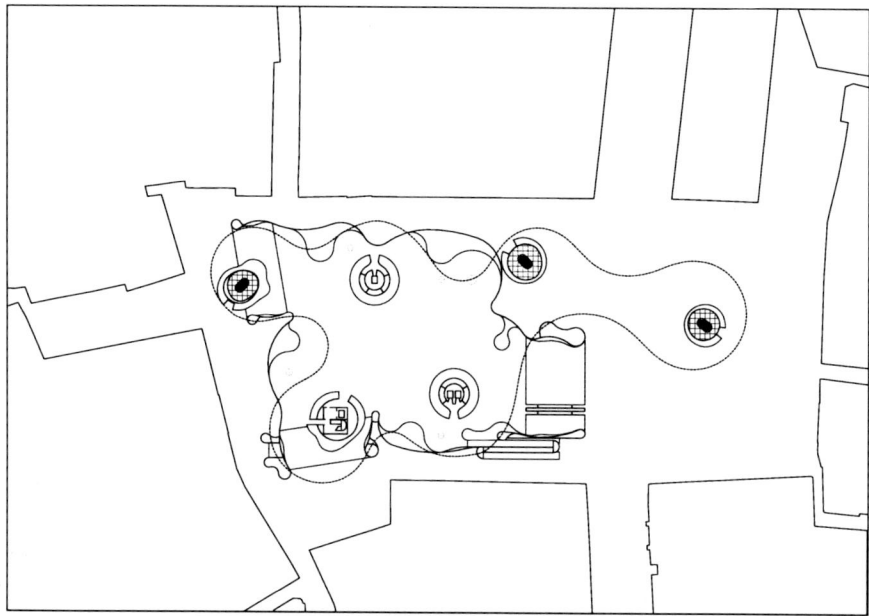

Fig. 6.14   The public plaza 16 ft 4 55/64 in (5 m) above ground

Fig. 6.15   The balcony plaza at 70 ft 6 15/32 in (21.5 m) above ground and the 'architectural walkway'

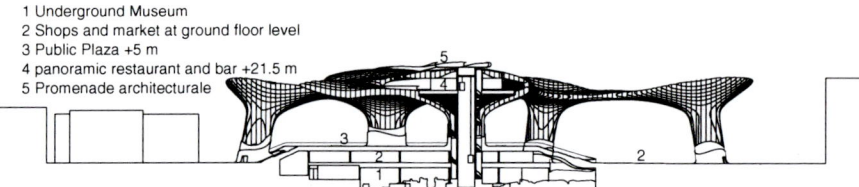

1 Underground Museum
2 Shops and market at ground floor level
3 Public Plaza +5 m
4 panoramic restaurant and bar +21.5 m
5 Promenade architecturale

Fig. 6.16 Longitudinal cross-section

Fig. 6.17 The city level with the shops and the market and the raised public plaza at 16 ft 4 55/64 in (5 m) above ground

(2007) expressly states that the project intends to create a relationship between its own forest and the other stone forest which is Seville Cathedral. The call is for a dialogue between them, a dialogue made of differences within a common identity. Gothic architecture is modelled on the branches of the trees united at the highest point: a physical form of stretching up to heaven, the spiritual space. This certainty in gothic architecture becomes clarity of spaces, structures, shapes and sizes. The space to which Metropol Parasol belongs is the public space of culture, art, commerce and relaxation, which had been provided in the historic city by central squares which were often opposite churches (Figs. 6.18 and 6.19).

Fig. 6.18 The nave of
Seville Cathedral, the third
largest of the Christian world

In the intentional parallel between the two forests, it is significant how Jürgen
Mayer H. aims to create the shape according to philosophical and aesthetic principles
of mimesis. Architecture has constantly re-worked references to animal and plant
shapes. The "column decorated with vines present in Western tradition since Greek
architecture is derived from the symbolism of the tree, and develops into a twisted
column thus acquiring similarities with the spiral shape" (Portoghesi 1999, p. 234),
with the geometric stylisation of the tree trunks and foliage in the Gothic style.

Contemporary architecture often finds justification in this process for developing
models of organic forms even further in order to fit its own ends and criteria
(Rykwert 1981; Visentin 2003; Pigafetta 2007).

In the modern reworking, that wide number of allegorical references, which
made a palimpsest of every ancient architectural structure, has been almost completely lost because it has been forgotten. The current interpretation is mainly
analogical: it extends some symbolic properties from one noted and definite reference to another through resemblance.

Fig. 6.19   Villard de
Honnecourt elevation from
the western tower of Laon
Cathedral, from the *Livre de
portraiture*, 1220–1235

In the design of the Metropol Parasol, the interpretation of the idea of a forest
proceeded from an initial idea of a structure in tubular steel covered by a wire
net. Although structurally and constructively effective, it was abandoned because
of the separation it entails in its composition between structure and cladding.
The design team, including the Ove Arup office in Berlin, agreed on the use of
engineered wood, which increases the mimetic value of the work and at the same
time satisfies the mechanical requirements and also the needs of environmental
sustainability.

The unitary freeform structure is superimposed on a grid of squares with sides
of length 4 ft 11 in (1.5 m). In the language of digital solid modelling, i.e. Boolean
logic, the result is obtained by subtraction between the two forms. The outcome
of this process constitutes the design for the second phase of the competition,
which was then built (Fig. 6.20).

The great forest of wood looms up larger than life, becoming a conspicuous point
on the city skyline. Precisely

as happens in nature where the flora and fauna develop in the places offering favourable
conditions, Metropol Parasol aims primarily to make summer temperatures more pleas-

Fig. 6.20   The grid of the parasol was generated on the digital model by removing from the organic forest a lattice of squares of 4 ft 11 in (1.5 m) side

ant, so as to allow services to be introduced and social life to nestle at its feet (Naboni 2011, p. 56)

The idea of the design is to create a huge sunshade—literally a parasol, hence the name—which should create in a natural way an urban micro-climate favourable to the activities taking place in its shade.

> If we are traditionally used to think of the sunscreening system as an addition to the building which contributes to improving its performance in selected ways, this design completely overturns this relationship (Naboni 2011).

In the design, each primary functional and architectural need is tackled according to the principle of coherence between the construction techniques and the expression of form. Thus, the base and the "trunks" of the trees which support the activities are cast in concrete. The restaurant platform is 70 ft 6 15/32 in (21.5 m) high and is made of steel. Also in steel are the ceiling beams of the underground museum so as to leave the maximum span of pillarless space for the exhibitions. The parasol is cut from composite timber panels with layers of veneer placed lengthwise and crosswise to the length of the panel (Fig. 6.21).

In order to calculate the dimensions of the single composite timber panels of the parasol, Ove Arup had to address the problem of interdependence between three variables for each element, namely: the thickness of the panel, the combined mass of the timber and the metallic connector and the forces acting at each node. The geometry of the structure depends on the forces acting on each node, and these forces in turn derive from the width and therefore the weight of the structural elements as well as from the

Fig. 6.21 The steel composite structure of the restaurant platform at 70 ft 6 15/32 in (21.5 m) above ground level

sizes of the necessary connectors. Since the structure in its entirety is composed of over 3,000 items, it was impossible to calculate directly the condition of equilibrium. Instead it was determined by successive approximations, with an iterative algorithm which varied the three variables of each item and evaluated over the whole structure whether the solution was getting closer to or further away from satisfying the load and geometric goals. The cycle was repeated until all the requirements were met. Because of the large number of elements, the process of convergence towards the condition of equilibrium took several days to compute (Koppitz et al. 2012, p. 254).

The resulting data on panel width and on the dimensions of the metal connector were integrated into the three-dimensional digital model of the entire structure of the parasol. This model was imported from Finnforest Merk, the timber contractor, in the CAD/CAM software system bocad-3D. From the 3D model of the structure, the system automatically worked out the NC program for controlling the Kuka robot with seven axes. For the work of each panel, the anthropomorphic robot carried out the unified manufacturing processes of cutting, milling and drilling. The manufacturing processes were assisted by new functions in bocad-3D for the control of imperfections in the shapes and for the creation of multiple connection joints between several elements (Fig. 6.22).

Calculation of the sizes of the composite timber items established variable panel widths of between 2 in 11/16 and 12 in 1/4 (68 and 311 mm), with a height of up to 9 ft 10 in (3 m). Because of the constraints of road transport from the Finnforest Merk factory in Aichach to the construction site, the maximum length of the beams was constrained to 49 ft (15 m). Some "special pieces", needed for construction of the "trunks" measured 54 ft 1 in 39/64 by 11 ft 5 in 51/64 by 5 in 33/64 (16.5 by 3.5 by 0.14 m). Altogether, the structure of the parasol was built of over 3,400 elements, with a total volume of about 4578 yd^3 (3,500 m^3) of composite timber (Schmid et al. 2011, pp. 707–714).

Fig. 6.22 Kuka robot
with seven axes, used
by Finnforest Merk
for the cutting, milling
and drilling of each panel

To optimise the use of composite timber and to minimize wastage, a custom-made
macro program calculated the best distribution and placement of each of the 3,400
items within the usable area of the plates. The panels were produced from Kerto-Q
composite timber with width varying between 61/64 in and 1 in 5/16 (24 and 33 mm),
glued together and consolidated in an autoclave in order to obtain the widths dictated
by the design. The panels were also drilled on the shop floor to make the bores
needed for the insertion of the connective rods for the nodes.

The design of the structural linking joints was developed collaboratively by Arup
and Finnforest. Generating the geometry of the grid ensures that all the connections
are at right angles, but each angle of connection between the panels is different. The
joints needed to work in a wide range of angles to guarantee stability between the
panels, to be easy to assemble on site, to accommodate tolerances in the construction phase and to be compact, since they are on view (Figs. 6.23 and 6.24).

The joint designed and made by Finnforest is made of a special fork-headed
connection. The flange of the fork is fixed to the panels using preloaded bolts that
are screwed into the rods inserted in the panels. It is a process similar to that used in
metal constructions.[7] On the shop floor, the threaded rods are fixed with glue into
the factory-drilled holes (25 19/32–27 9/16 in deep, 65–70 cm) in the ends of the

---

[7] See Sect. 4.4.

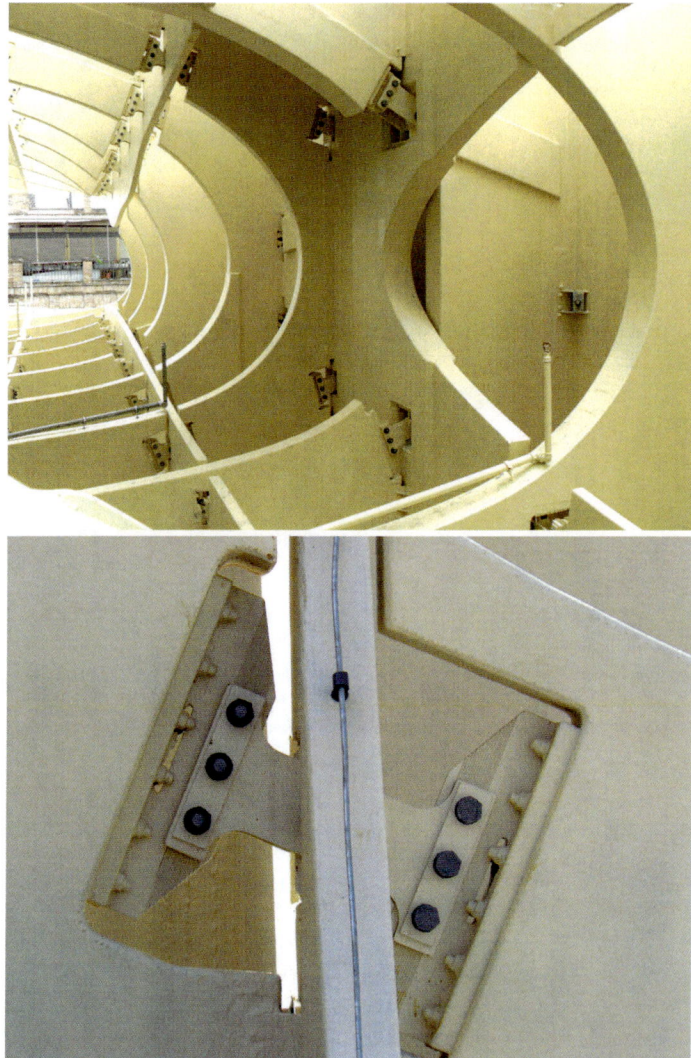

Figs. 6.23 and 6.24 The nodes at the joining points of the structure must function in a wide range of angles, guarantee stability between the panels and be compact as they are on view

panels. The purpose of the rods is to transfer the loads and stresses to a broader resistant section of the composite timber (Fig. 6.25).

On site, the panels are lifted in the assembly position, where the fork-headed hooks are fixed in their final position with the insertion of steel pins. The pins give stability by transferring axial forces and flexible moments. In this way, the mounting process is simplified and standardised and flexible at the same time because it can adapt to the multiplex load and connection conditions of the large structure (Figs. 6.26, 6.27, and 6.28).

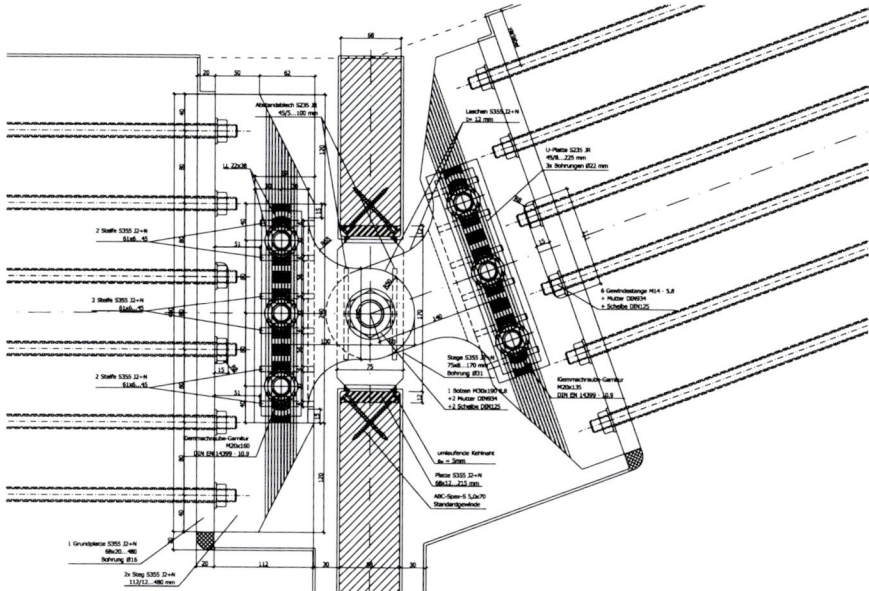

Fig. 6.25 Detail of the structural node made of the fork-headed connection. The coupling hooks are fixed to each other with steel pins. The flange of the fork is fixed to the panels using preloaded bolts screwed into the threaded rods

Fig. 6.26 The threaded rods are fixed to the panel by gluing in the holes made in the panel end

Fig. 6.27 The special fork-headed connection is bolted onto the rods of the panels

Fig. 6.28 The panel is lifted into the assembly position, where the fork-headed hooks are fixed in their final position with the insertion of steel pins

6.5   The Rebuilding of the Inner Dome of the Teatro Petruzzelli in Bari[8]

6.5.1   History of the Rebuilding

On December 6th, 2008, reconstruction of Bari's Teatro Petruzzelli was completed. Since its destruction due to an arson attack in 1991, the theatre had taken 18 years to rebuild, not because of technical problems so much as because of disputes. Legal wrangles set the family, who own the opera house, against Bari's town council, who own the land on which the building stands, and in the final analysis set them both against the interests of the citizens hoping for a swift reconstruction of the theatre.

> From the start it was clear that the wish of all the citizens of Bari was that the opera house be rebuilt 'as it was', and the first funds, given by Italy's central government immediately after the fire, be used to rebuild the external dome and the roofs exactly as they were before the fire (Blasi 2009).

Work of stabilisation and recovery resumed almost immediately after the fire, only to be stopped and then started again when owners and management brought legal actions against each other. In this period, designs for recovery were commissioned by the proprietors from trusted designers and later by the Superintendent from the National Environment and Architectural Heritage Body.[9] It is the designs of the latter organisation which were ultimately built.

The first funding, given by central government immediately after the fire, was used to rebuild the large steel external dome, about 131 ft 2 in 13/16 (40 m) in diameter and 52 ft 5 in (16 m) high, which soars above the auditorium of the theatre.

> It is a recurring and understandable phenomenon that when highly symbolic buildings are lost through traumatic events such as collapses, fires, earthquakes or wars, the will of the people is to see the work rebuilt as much as possible exactly as before, as if to lay the ghost of the destructive action (Blasi 2009).

The external dome and the roof were rebuilt as they were before the fire, preserving unaltered the original geometry and structural plan.

The external dome and roof defined the size and structural restrictions of the reconstruction project. The design aimed to use modern technologies to rebuild the inner dome, which was originally built in timber and wattle-and-daub. The new inner dome is still made of timber, but with the advantages of glulam. Turning to a glue-laminated timber system allowed the designers to use thinner widths, which are more efficient from the point of view of stability and consequently help increase the load bearing capacity for the equivalent span. Compared with the original structure in solid timber, the laminated timber ensures constant mechanical resistances, consistent quality and greater static efficiency (Fig. 6.29).

---

[8] Text and images by the architect Alberto La Tegola, consultant at Stratex S.p.A.

[9] Beni Architettonici e Ambientali.

Fig. 6.29 Plan of the timber
structure of the inner dome

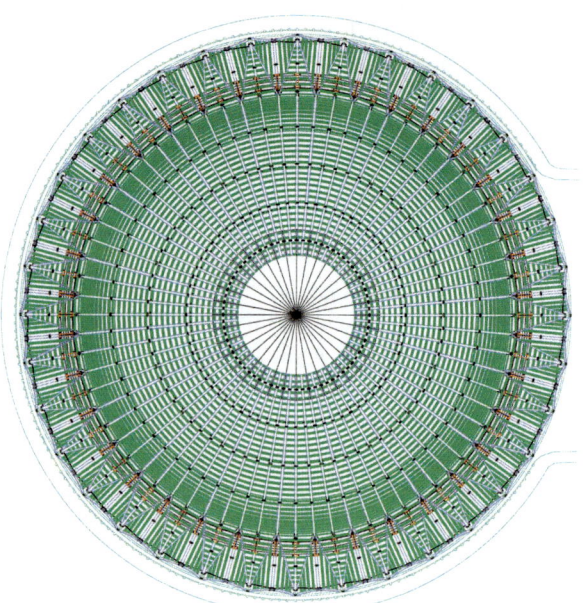

   The designers decided to make use of the building opportunity offered by prefabrication by making technological units and elements in the factory and later assembling them in a highly organised building site. The design choices, finalised in the technological and building system, satisfied the stringent requirements of managing the weight of the entire structure; the acoustic characteristics; matching the geometry of the pre-existing external dome; the construction schedule and costs.

6.5.2 Construction Project

From the tender documents, the "Petruzzelli Reconstruction" consortium of contractors launched the bid for the glulam supplier, which led to the company Stratex S.p.A. being commissioned to provide all the wooden structural works for the internal dome and the supporting structure of the auditorium.

   Stratex technical staff developed the construction project according to the lines defined by the designers, enabling the erection of a great structure within the constraints imposed by the pre-existent outer dome.

   The starting point for the construction project was the survey, carried out and provided by the Consortium, and the bid documentation.

   The Stratex team conceived the prefabrication of the technological elements in the factory. These elements were then to be assembled later on site, putting together the complex structure according to rationalised time and cost procedures.

   One specific constraint and challenge for the management of the construction process was handling the parts to and into the building site. The entrances to the

Fig. 6.30 The rods which join the metal structure of the outer dome and the timber structure of the inner dome with the interposition of acoustic insulators

theatre auditorium from the street level have a height restriction of only 8 ft 2 7/16 in (2.5 m). It was also necessary to face the problem of lifting the prefabricated parts up to the area where the inner dome was to be assembled, 65 ft 7 13/32 in (20 m) above the floor of the auditorium. In the construction design phase, these constraints were addressed by setting the dimensions of the prefabricated parts to fit the logistics of the building site.

Having settled upon the dimensions of the elements, the design team addressed the correct structural, acoustic and spatial interaction between the underside of the outer steel dome with its load bearing function and the internal dome of laminated timber. The structural relationship between the two domes, external and internal, was achieved with tie rods (Fig. 6.30).

The inner dome was assembled in quadrants, each quadrant being raised into place after the glulam elements had been assembled within the footprint of the dome. The parts were fixed together and to the brick drum with flanges and clamping bolts. Each quadrant was lifted into the designed position and hung with tie rods, which were anchored to the load-bearing outer dome. Acoustic insulators had been placed inside the tie rods in order to soften vibrations and resulting resonances. This optimised the acoustic performance of the dome (Fig. 6.31).

Fig. 6.31 The erection of the inner dome in laminated timber

6.5.3   Three-Dimensional Modelling of the Structure

The two-dimensional surveys of the steel outer dome, in AutoCAD format, were imported into the CadWorks modelling system used by Stratex. The surveys, updated and verified with sections surveyed on the building site, defined the basic measurements for the three-dimensional model of the already-built steel dome. This model was the geometric and dimensional reference for modelling the glulam structure.

Three-dimensional modelling allowed Stratex to optimise the configuration of the new structure as regards the effective insertion and integration between the two domes, the steel outer dome and the glulam inner dome. This methodology allowed them to "match" the closure of the counter-wooden dome at the porthole windows, restoring the ornaments according to their original geometry. With this method, the problem of interference between the beams of the steel structure and the supports of the lunettes in wood was also resolved (Figs. 6.32 and 6.33).

The three dimensional model of the design, inserted into the exact model of the existing building, allowed a process of "virtual" checking. The analysis process aimed to discover the maximum possible clearances, necessary to achieve the geometry being designed, and to assess the statics aspects.

Fig. 6.32 Three-dimensional
model of the laminated timber
inner dome, the steel outer
dome and the brick drum

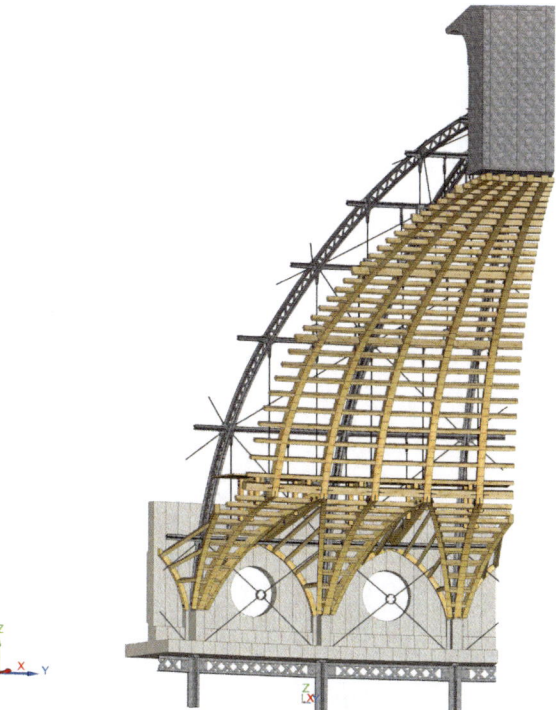

6.5.4   *Structural Check*

The design team sensibly took advantage of the given constraints and simplified the
force diagram and the structural shape, defining the maximum widths of the timber
elements according to the maximum possible clearances. The number and the
placing of the anchorages for the connecting rods between the two domes were
considered integral parts of the force diagram, being responsible for the fastening
and consolidation of the glulam and steel domes.

The sizing of the suspensions and consequently the calculation of the reactions
transferred made it possible to quantify and calibrate the sound-absorbing insulating
material inserted in the suspensions in order to dissipate any transferable resonance
or vibration from the sound environment of the theatre to the steel dome through the
timber inner dome.

Also regarding the insulators, the regulatory requirement of inserting seismic
insulators was also considered in the dynamic design of structure. This was undertaken in accordance with the restoration philosophy which, as far as possible, aims
to "improve" performance whenever it is impossible to fulfil the current standards
of earthquake resistance. Finally, the role played by the "virtual checking" process
in planning the handling of the glulam elements within the building must not
underestimated.

Fig. 6.33 Three-dimensional model showing the tie roads between the internal and external domes

The outcome of the process is a structure in laminated timber, composed altogether of 4,820 glulam elements, straight and curved, of length varying between 23 in 5/8 and 630 in (60 and 1,600 cm): 72 curved beams in light laminated timber,[10] primary and secondary framework, 108 lower ribs in light laminated timber, a few thousand moulded glulam planks, inspection planks and 53/64 in (21 mm) wide perforated board.

6.5.5 Prefabrication on the Shop Floor

For the Stratex factory, manufacturing a building with a coverage of 10,872 ft^2 (1,010 m^2) is part of normal production capacity. The unusual aspect, if we want to

---

[10] Working with thin veneers or micro-veneers refers to beams composed of veneers less than 1 in 37/64 (4 cm) thick. Such veneers are used for curved elements with a high degree of curvature. Their depth, in relation to the radial of curvature, must be S < R/200.

Fig. 6.34   Detail of the carpentry work of the joint between round windows and beams

find something extraordinary, comes from the number and width of the components. Using 4,820 pieces of laminated timber to cover a site with an area of around 10,764 ft^2 (1,000 m^2) is highly unusual when we consider that similar structures, of a similar coverage, involve less than a hundred pieces. As for the widths, on the other hand, these are "small fry" for the numerically controlled production process used by Stratex.

Stratex uses a proprietary CAD/CAM system developed in CadWorks to automate the generation of the structure's pieces in accordance with the requirements of the design. In the case of the Petruzzelli building, the system generated the numerical control for the various unit processes of the production that were necessary to produce the finished pieces in a condition ready for assembly. Each piece of the structure, after the lamination processes of the beams or timber elements, proceeds automatically through the various processes of moulding, rubbing down, pre-drilling, matching and finishing. Stratex's experience has helped to work out the critical node problems of coupling the elements with each other and with the building: the integrated design took into account the whole process including steel joints, matching holes for them and the necessary means of fixing them into the holes (Fig. 6.34).

Each piece of the building is countersigned with a unique code which identifies the individual piece from the automatic production sequence to the building site. In construction, the code is used to place the piece in the exact coordinates defined in the three-dimensional building model and for subsequent checks.

One critical element of the production phase depends on the system of fixing the piece in the work centre. In order automatically to hold the piece in the working position, the work station uses vacuum clamping, calibrated up to a minimum limit of dimensions and piece volume. Many pieces of the building turned out to be below this limit. Stratex made a custom-built template for scarfing, cut from a thick beam.

Scarfing the piece to be worked on to the beam made it possible to create a solid system which the sensors of the machining centre classed as within the minimum requirements for the locking, which was useful for working in normal automatic mode.

All the pieces of the building were manufactured in 3 weeks (Fig. 6.35).

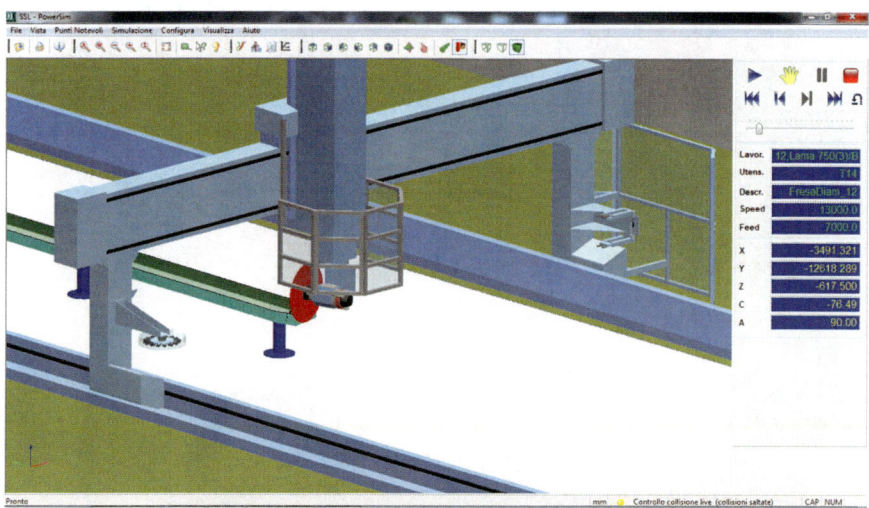

Fig. 6.35 Milling of beams in laminated timber through work centre in the Stratex plant

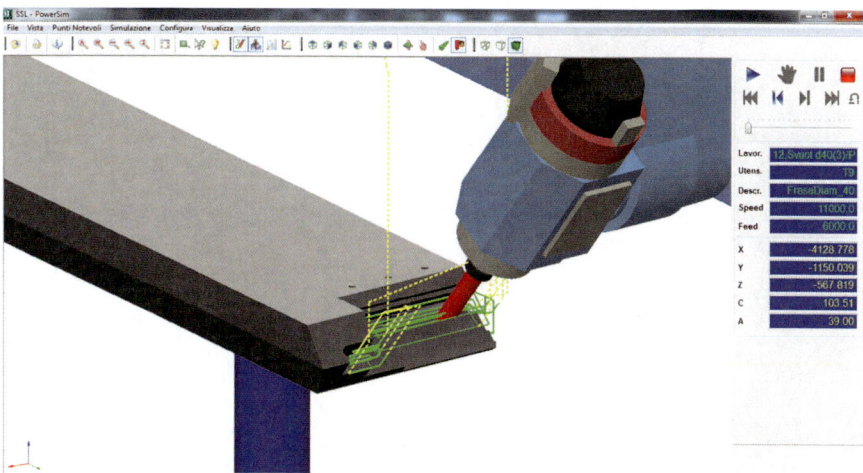

Fig. 6.35 (continued)

6.5.6 Assembling on Site

The integrated process of design and manufacture, based on the accurate three-dimensional modelling of the dome inserted in the model of the steel dome and the perimeter wall, assisted in the prefabrication of the elements whose dimensions and weight had to conform to the access restrictions of the building site area and to the means of handling suitable to the specific construction area.

Fig. 6.36 One of the 72 curved main beams, 16 m long at the level of the building site

The material was first raised to the level of the roof of the theatre with a crane. It was then introduced manually through the round windows at the base of the dome itself.

The process of "virtual checking" confirmed its effectiveness, allowing even the 72 main curved beams, 52 ft 5 in 59/64 long (16 m), to pass through the round windows. These were assembled element by element in pairs braced and panelled then hoisted up to the level of the upper lantern using a manual block-and-tackle hoist. The erection process was planned so as never to exceed the precisely calculated maximum loads for the steel dome (Fig. 6.36).

The erection work, which took 3 weeks, did not encounter a single hitch that could be attributed to inaccuracies in the dimensions of the parts. It must be stressed that, at the moment of "closing the circle", in other words when the last couple of beams was put into place, in fact closing the circumference of the base of the dome (the length of the circumference is a little less than 328 ft (100 m) the error encountered was hardly 1 in 21/32 (42 mm), a negligible tolerance in timber construction (Figs. 6.37 and 6.38).

Fig. 6.37   Detail of the laminated timber structure

Fig. 6.38   The structure of the inner dome completed before being covered

References

Blasi C (2009) La nuova cupola del teatro Petruzzelli. L'ingegneria edilizia ambiente e territorio 23–24(4):60–65

Callegari G, Zanuttini R (2010) Per una promozione sostenibile della filiera legno in edilizia. In: Callegari G, Zanuttini R (eds) Boislab. Il legno per un'architettura sostenibile. Alinea, Firenze

Cremonini C, Zanuttini R (2010) I prodotti della prima lavorazione del legno. In: Callegari G, Zanuttini R (eds) Boislab. Il legno per un'architettura sostenibile. Alinea, Firenze

Giordano G, Ceccotti A, Uzielli L (1999) Tecnica delle costruzioni in legno. Caratteristiche, qualificazione e normazione dei legnami da costruzione; progettazione e controllo delle strutture lignee tradizionali; applicazione dei moderni metodi di calcolo alle nuove tipologie costruttive; classi di resistenza dei legnami strutturali secondo la normativa europea. Hoepli, Milano

Hoadley B (2000) Understanding wood. A craftsman's guide to wood technology. Taunton Press, Newtown

Koppitz JP, Quinn G, Schmid V, Thurik A (2012) Metropol parasol – digital timber design. In: Gengnagel C, Kilian A, Palz N, Scheurer F (eds) Computational design modeling. Springer, Berlin

Lehmann WF (2011) Wood-Based Composites and Laminates. In: Kroschwitz JI, Seidel A (eds) Kirk-Othmer encyclopedia of chemical technology. Wiley-Interscience, Hoboken

Massobrio G, Portoghesi P (1990) Casa Thonet. Storia dei mobili in legno curvato, 2nd edn. Laterza, Bari

Mayer JH (2007) Sleeping beauty: Potientiale in Architektur und Ästhetik. In: Workshop form und Funktion: Zur Frage der Qualität in der Architektur, Wolfsberg Arts Forum, Ermatingen, 15–16 Mai 2007

Naboni E (2011) Soft performance nei sistemi di schermatura. Tenda in out 2:56

Pigafetta G (2007) Storia dell'architettura moderna. Imitazione e invenzione fra XV e XX secolo. Bollati Boringhieri, Torino

Portoghesi P (1999) Natura e architettura. Skira, Milano

Rosenberg N (1976) Perspectives on technology. Cambridge University Press, Cambridge

Rykwert J (1981) On Adam's House in Paradise. The idea of the primitive hut in architectural history. MIT Press, Cambridge, MA

Sala N, Sala M (2005) Geometrie del design. Forme e materiali per il progetto. FrancoAngeli, Milano

Schmid V, Koppitz JP, Thurik A (2011) Neue Konzepte im Holzbau mit Furnierschichtholz – Die Holztragkonstruktion des Metropol Parasol in Sevilla. Bautechnik 88:707–714

Sinopoli N, Tatano V (eds) (2002) Sulle tracce dell'innovazione. Tra tecniche e architettura. FrancoAngeli, Milano

Tampone G (2011) Strutture e costruzioni autarchiche di legno in Italia e Colonie. Caratteri e criteri di conservazione. Boll Ing 11:3

Thomson R (2009) Structures of change in the mechanical age. Technological innovation in the United States 1790–1865. Johns Hopkins University Press, Baltimore

Visentin C (2003) L'equivoco dell'eclettismo. Imitazione e memoria in architettura. Pendragon, Bologna

Project Credits

4.4 Foster + Partners, Swiss Re Tower, London

Architect: Foster and Partners
Structural Engineering: Ove Arup & Partners
Mechanical & Electrical Engineers: Hilson Moran Partnership Ltd.
Project Management: RWG Associates Ltd
Planning: Montagu Evans \ Richard Coleman Consultancy
Quantity Surveyors: Gardiner & Theobald
Interior Designer: Benett Interior Design
Lighting Consultant: Speirs and Major Associates
Acoustics Consultant: Sandy Brown Associates LLP
Elevator Consultant: Van Deusen & Associates
Facade Consultant: Emmer Pfenninger Partner AG
Security System Consultant: Videf Security Management Ltd.
Wind Surveyor: Rowan Williams Davies & Irwin Inc.
Developer: Swiss Re
General Contractor: Skanska UK
Demolition: Keltbray Ltd
Piling: Cementation Foundations Skanska Ltd
Real-estate Agent: DTZ Debenham Tie Leung
Elevator Supplier: KONE
Facade Maintenance System Suppliers: Lalesse Gevelliften BV, B. Teupen Maschinenbau GmbH
Facade Supplier: Schmidlin Ltd.
Sub- and Super-structure: PC Harrington Contractors Ltd
Top of building: Waagner Biro Ltd
Steel Suppliers: Hollandia BV, Victor Buyck Steel Construction NV
Cleaning equipment: Street Craneexpress Ltd
Legal advisers: Linklaters & Alliance

L. Caneparo, *Digital Fabrication in Architecture, Engineering and Construction*,
DOI 10.1007/978-94-007-7137-6, © Springer Science+Business Media Dordrecht 2014

5.6   Barkow and Leibinger, Gatehouse, Stuttgart

Architect: Frank Barkow and Regine Leibinger, Barkow Leibinger Architects
Design Team: Meredith Atkinson, Carsten Krafft
Construction Team: Caspar Hoesch
Structural Engineering: Werner Sobek Ingenieure
Project management: Gassmann + Grossmann
Landscape: Büro Kiefer
Façade: Ove Arup Berlin, Werner Sobek Ingenieure

5.7   Randall Stout Architects, Art Gallery, Alberta

Architect: Randall Stout, Randall Stout Architects
Architectural technologist: Manzer Mirkar, Randall Stout Architects
Construction management: Trevor Messal, Ledcor Construction Ltd
Structural Engineering: DeSimone Consulting, BPTEC-DMW Engineering Ltd
Mechanical & Electrical Engineers: IBE Consulting Engineers, Stantec Consulting Ltd
Structural Steel: Empire Iron Works Ltd
Metal/Cladding Supply: A. Zahner Company
Architectural Metal/Cladding: Flynn Canada Ltd
Curtainwall/Envelop: Reed Jones Christoffersen Ltd
MEP/Fire Protection: Stantec Consulting Ltd

6.4   Jürgen Mayer H. Architects, Metropol Parasol, Seville

Architect: Jürgen Mayer H., Andre Santer, Marta Ramírez Iglesias
Design Team: Ana Alonso de la Varga, Jan-Christoph Stockebrand, Marcus Blum,
    Paul Angelier, Hans Schneider, Thorsten Blatter, Wilko Hoffmann, Claudia
    Marcinowski, Sebastian Finckh, Alessandra Raponi, Olivier Jacques, Nai Huei
    Wang, Dirk Blomeyer
Construction Engineering: Ove Arup
Structural Team: Romain Buffat, Kayin Dawoodi, Steffen Janitz, Andres Garzón,
    Enrique González, Carlos Merino, Estrella Morato, Victor Rodríguez Izquierdo,
    Volker Schmid
Project Management: Jose de la Peña, Jan-Peter Koppitz
Mechanical & Electrical Team: Salvador Castilla, Alborada Delgado, Marta
    Figueruelo Calvo
Fire Protection Team: George Faller, Benjamin Barry-Otsoa, Jimmy Jonsson
Timber engineering and detail design: Finnforest, Aichach

6.5   The Rebuilding of the Inner Dome
of the Teatro Petruzzelli, Bari

Construction Project Coordinator: Carlo Blasi
Construction Team: Consorzio ATP per il Petruzzelli
Project Coordination: Soprintendenza per i Beni Ambientali, Architettonici, Artistici
    e Storici della Puglia
Architect: COMES Srl, Luigi Sylos Labini, Studio Associato di Architettura SMN
Structural Engineering: Amedeo Vitone, Studio Vitone Associati
Wood Construction: Stratex SpA
Wood Construction Management: Alberto La Tegola

Illustration Credits

Chapter 2

Fig. 2.1 McBean House built with the Marshall Erdman prefabrication system from a design by Frank Lloyd Wright

Fig. 2.2 The first machine for horizontal milling produced by Pratt & Whitney in about 1860 (Pratt & Whitney Inc)

Fig. 2.3 Operation of the Gunnison Magic Homes factory. (**a**) Sawing the plywood. (**b**) Sawing the uprights of the structure. (**c**) Assembling the panels. (**d**) Trimming the panels. (**e**) Multi-function press for gluing panels together. (**f**) Panel painting line (From Burnham Kelly (1951) The prefabrication of houses. A study of the Albert Farwell Bemis Foundation of the prefabrication industry in the United States. MIT Press, John Wiley and Sons, Cambridge, MA, New York)

Fig. 2.4 The balloon frame construction system (From American Wood Council (2001) Details for Conventional Wood Frame Construction. National Forest Products Association, Leesburg, VA)

Fig. 2.5 The platform frame construction system (From American Wood Council (2001) Details for Conventional Wood Frame Construction. National Forest Products Association, Leesburg, VA)

Fig. 2.6 Prototype Type 32 of Ferdinand Porsche's design for the Volksauto

Fig. 2.7 Precursor of the timber processing centre, Greenwich machine works, circa 1870

Fig. 2.8 Detail of the corner joint in the Liegender Blockbau

Fig. 2.9 Nordische Blockhäuser by Christoph & Unmack: detail of the wall and of the floor

Fig. 2.10 Sommerfeld House in Berlin was the first commission Gropius received as Director of the Bauhaus in 1920

Fig. 2.11 Ernst Wasmuth Verlag AG and Birkhäuser GmbH. General view from the road of the "director's house" in Niesky (From Wachsmann K (1930) Holzhausbau.

L. Caneparo, *Digital Fabrication in Architecture, Engineering and Construction*, DOI 10.1007/978-94-007-7137-6, © Springer Science+Business Media Dordrecht 2014

Technik und Gestaltung. Ernst Wasmuth Verlag AG, Berlin. New edition: (1995) Birkhauser, Basel)

Fig. 2.12 (**a**, **b**) Two, three or four panels may be assembled together using Y-shaped metal connectors

Fig. 2.13 United States Patent Office Patent no. 2,355,192, filed by Wachsmann and Gropius in 1942

Fig. 2.14 United States Patent Office Patent no. 2,421,305, filed by Wachsmann and Gropius in 1945

Fig. 2.15 The new patent connector made with flat parts, more robust and simpler to manufacture (Akademie der Künste, Berlin, Konrad Wachsmann Archive, KWA-01-99-F.8a, KWA-01-99-F.9a, photographer Anna Wachsmann)

Fig. 2.16 Wachsmann diagrams for modular coordination in two-dimensional space, three-dimensional space and in the time dimension, for the assembly (Reinhold Publishing Corporation)

Fig. 2.17 Isometric view of the General Panel Corporation plant in Burbank. "A production line. Symbol of the concept of automation. A certain number of machine tools, mechanically and chronologically synchronised and servo-driven, shape the materials into the finished product" (From Wachsmann K (1961) *Turning Point of Building: Structure and Design*. Reinhold Pub. Corp., New York. Akademie der Künste, Berlin, Konrad Wachsmann Archive, KWA-0001-093-F.06)

Fig. 2.18 Sheets of Douglas fir plywood are glued into the frames in only 5 s with special high-frequency machines. An electronic press consolidates the entire stressed skin panel in a single operation. (Akademie der Künste, Berlin, Konrad Wachsmann Archive, KWA-0001-096-F.01a)

Fig. 2.19 Working prototype of anthropomorphic robot with seven degrees of freedom, made by Wachsmann in his Laboratory at the University of California towards the end of the 1960s (Akademie der Künste, Berlin, Konrad Wachsmann Archive, KWA-0001-169-F-81)

Fig. 2.20 Hydraulic sawmill in America at the beginning of the nineteenth century (Barbara and Roger Bcrkt)

Fig. 2.21 (**a**, **b**) LaBelle Works factory in Wheeling, Ohio, for the industrial production of nails using machines for cutting and manufacturing the metal wire (Ohio County Public Library)

Chapter 3

Fig. 3.1 Colossus, the first digital computer in history, operating at Bletchley Park in 1944

Fig. 3.2 Electronic Numerical Integrator And Computer (ENIAC) introduced in 1946. Goldstine estimated the cost of the project to be the equivalent of five million dollars in 1970

Fig. 3.3 Card-a-matic Milling Machine at the Servomechanism Laboratory of Massachusetts Institute of Technology (MIT Museum)

Fig. 3.4 Example of the APT II program. Symbolic names are assigned to the geometric primitives: JIM is a point with coordinates 6 and 7, JANE is the circumference of a circle of radius 2 and centre JIM, WALDO is the outline obtained by a tangent to the circles JANE and JUNE

Fig. 3.5 The illustration from 1910, entitled *Vision of year 2000*, shows how the progress was imagined: the architect from his operator panel has full and direct control of the construction process (Bibliothèque nationale de France)

Fig. 3.6 Robotised assembling of the bricks for the walls of Gantenbein Vineyard. The elements of the façade were fabricated from standard bricks. The industrial robot has applied the adhesive and positioned each brick (Gramazio & Kohler, ETH Zurich)

Fig. 3.7 The number of industrial districts by region, monitored by the Italian National Centre

Chapter 4

Fig. 4.1 Profile milling (contouring) with circular plates (Dr. Johannes Heidenhain GmbH)

Fig. 4.2 Mechanised plant after the operations of casting for the production of casts between 50 and 500 kg, also for small series (Fonderie Ariotti SpA)

Fig. 4.3 Plasma processing plant, using partially or totally ionised gas, in which neutral molecules, positive ions and free electrons are present at the same time (Kolzer Srl)

Fig. 4.4 Deposition process using the technique of PECVD (Plasma Enhanced Chemical Vapour Deposition), based on the excitation and ionisation of silicon (SiOx). It creates a nanostructured polymer film. The polymer is deposited on the surface of the wood and becomes all in one with it. The film has a chemical composition similar to quartz (with a high elasticity), and makes the surface of the wood water-repellent, protects from chemicals and organic solvents, and resists UV light. After the ionic deposition only one coat of varnish is required (Kolzer Srl)

Fig. 4.5 Wood samples after 6 years of exposure to the elements: on the *left* treated with PECVD, on the *right* not treated (Kolzer Srl)

Fig. 4.6 Processes of slicing and moulding through electronically-fed press (OMR Snc di Giacomini)

Fig. 4.7 Hydraulic bending press controlled by CNC to carry out bending of semi-worked metal pieces. The press is controlled by a robot, which carries out manipulation, loading and unloading of the pieces (OMR Snc di Giacomini)

Fig. 4.8 Panelling area at OMR Snc, which carries out all the operations of bending of a punched and cut piece so as to transform it into a finished component. The loading and all the bending operations with the necessary rotations and the relative positioning are automatic. Finally the component is removed from the bending centre by the robot which manages the palletisation (OMR Snc di Giacomini)

Fig. 4.9 Robotised work areas for pressing-bending and for panelling are useful in making medium-large series: single sheet doors from OMR Snc (OMR Snc di Giacomini)

Fig. 4.10 Bead sintering process for the production of diamond thread, used in cutting stones (THEBEAD)

Fig. 4.11 Mixing and pressing of the beads (THEBEAD)

Fig. 4.12 Sintering furnace (THEBEAD)

Fig. 4.13 The finished beads, ready for the assembly of the diamond wire (THEBEAD)

Fig. 4.14 View of the UTS Distort Pavilion, an educational project at the Sydney Faculty of Architecture (Distort Pavilion, 1st year B.Des.(Arch) student elective group/Joanne Jakovich/Melonie Bayl-Smith, University of Technology Sydney)

Fig. 4.15 The erected pavilion with the group which participated in the project (Distort Pavilion, 1st year B.Des.(Arch) student elective group/Joanne Jakovich/ Melonie Bayl-Smith, University of Technology Sydney)

Figs. 4.16 and 4.17 Prototyping of the structural nodes (Distort Pavilion, 1st year B.Des.(Arch) student elective group/Joanne Jakovich/Melonie Bayl-Smith, University of Technology Sydney)

Figs. 4.18, 4.19, and 4.20 D-Shape is a rapid prototyping system for making large stone-like objects, on the scale of buildings. Successive layers of sand 13/64 in (5 mm) thick are deposited. By means of an inorganic binder, the sand sets in a microcrystalline structure, making a stone-like conglomerate with a high resistance to traction (DINITECH SpA)

Figs. 4.21, 4.22, and 4.23 Manufacturing process for the Radiolaria pavilion, a project by architect Andrea Morgante of Shiro Studio. The shape is inspired by the amoeboid protozoa of the same name. The stages of prototyping: the mould of the monolithic structure; the removal of inert materials that were not consolidated, and which will be reused; the finished pavilion after finishing and polishing (DINITECH SpA)

Fig. 4.24 Shimizu Corp., the building under construction seems to be wearing a "top hat" (Shimizu Corp)

Fig. 4.25 The platform is lifted from one storey to the next by four jacks (Shimizu Corp)

Fig. 4.26 Shimizu Corp., the building site resembles a factory (Shimizu Corp)

Fig. 4.27 The whole construction process is supervised from a control cabin (Shimizu Corp)

Fig. 4.28 Robotised system for laying floors (Shimizu Corp)

Fig. 4.29 Frank Gehry's Fish Sculpture at the Olympic Village in Barcelona (Till Niermann)

Fig. 4.30 Gehry's physical model (Joshua White, JW Pictures Inc., courtesy Gehry Partners LLP)

Fig. 4.31 CATIA master model of the main structure of the sculpted fish shape (Rick Smith, virtual build technologies)

Fig. 4.32 CATIA model diagramming individual secondary structural pipe layout for bending fabrication (Rick Smith, virtual build technologies)

Fig. 4.33 CATIA model of the structural steel pipes with connections to primary structural steel (Rick Smith, virtual build technologies)

Fig. 4.34 Permasteelisa was given the job of the construction design and developed a complete construction methodology in which the skin shapes the inside. The skin defines the secondary and the primary structure (Attilius)

Fig. 4.35 A database keeps a record for each piece with a unique identification code, its place in the gantry and its progress (Permasteelisa Group)

Fig. 4.36 Rather than the blueprints, it is the database which manages and organises the progress of the project: it allows the tracking of each single part from manufacture to building site, and checking the work progress status (SpAragnamu)

Fig. 4.37 Foster + Partners. Swiss Re Tower plan of the 50th, 39th, 33rd, 21st and 6th floors and the entrance with the arrangement of the public spaces (Foster + Partners)

Fig. 4.38 The tower has a diameter of 40 m on the ground floor, and expands to 57 m in the central section (Foster + Partners)

Fig. 4.39 The tower is 180 m high on 50 floors (Aurelien Guichard Creative Commons)

Fig. 4.40 Foster + Partners interpret the needs of the commissioners in a magnificent organic form which has become a tourist attraction and a conspicuous part of the London skyline

Fig. 4.41 Cross-section view of the tower (Gregory Gibbon / Foster + Partners)

Fig. 4.42 Foster's sketch shows the process of generating the building around the hub: each floor is rotated by 5° compared with the one below. The circular movement between floors makes the light wells go round the building (Norman Foster)

Fig. 4.43 The building site shows the characteristic diagonal geometry of the networked structure, achieved with A-shaped frames connected to the horizontal hoops of tubular steel sections which encircle each floor (James Newman)

Fig. 4.44 The building site of the Swiss Re Tower; in the foreground is the node-linking hoops, columns, and the ties for the fixtures. Each piece is numbered sequentially so that it can be immediately identified and checked (Larry Sass MIT)

Fig. 4.45 The triangular tessellation of the reticular structure of the façade coincides with the windows obtained by juxtaposing two isosceles triangles in a diamond-shape (Foster + Partners)

Chapter 5

Fig. 5.1 Horizontal lathe with working area 8 in 21/32 by 13 in 25/64 (220 by 340 mm) (Takisawa Machine Tool Co)

Fig. 5.2 Milling with a five-axis machine with tilting worktop furnished with multiple rotating piece-holders (Dr. Johannes Heidenhain GmbH)

Fig. 5.3 Grinding a dental crown (Dr. Johannes Heidenhain GmbH)

Fig. 5.4 Five-axis vertical machining centre with work area 26 in 3/8 by 32 in 9/32 by 23 in 5/8 (670 by 820 by 600 mm) (Buffalo Machinery Co)

Fig. 5.5 Frame in birch plywood, made with NC milling machine, for investment cast of 3 MW wind turbine (Fonderie Ariotti SpA)

Fig. 5.6 Schnell Wire Systems cold-rolling lines for the production of smooth and ribbed wires for reinforced concrete (Schnell SpA)

Fig. 5.7 Cassette for rolling or profiling operations, consisting of three rollers bent at 120°. The rollers cause the reduction of the section and, on request, of the ribs in relief (Schnell SpA)

Fig. 5.8 Electronic control through Programmable Logic Controller. It manages the different parameters of working of the rolling machine: (1) set up or control of the wire feeding speed; the speed of the line depends on several factors e.g. the quality of the wire, lubricant, quality of the rollers, experience of the operators and mostly on the spooling unit; the motors are set to steady torque and power, each controlled by frequency converters; (2) set up of the final weight of the spooler according to the wire; (3) control of the pressure of the carriage device to prevent wire breakage while in tension; (4) electronic set up of the spacing between each layer of the spooler; (5) production detection; (6) maintenance detection (Schnell SpA)

Figs. 5.9 and 5.10 Welding of metallic enclosure with six-axis anthropomorphic robot, configured for arc welding, mounted on rail of 315 in (8,000 mm) length (Air Liquide Welding)

Figs. 5.11 and 5.12 Comau SMART-5 NJ4 anthropomorphic robot with six degrees of freedom and maximum payload at the wrist 170 kg. The wrist is hollow to contain all the wires. The kinematic structure is designed to reduce weight, inertia and interferences (Comau SpA)

Figs. 5.13 and 5.14 Barkow Leibinger Architects. Plan of the reception building, the entrance to the Trumpf factories. Longitudinal section of the reception building (Barkow Leibinger Architects)

Fig. 5.15 (a–c) The glass walls are made with a cavity of 10 in (25 cm), filled with acrylic glass cylinders (Barkow Leibinger Architects)

Fig. 5.16 (a, b) Girders assembled into a 24-in-thick (60 cm) honeycomb, load-bearing structure; the intermediate stiffeners contribute to stiffness of the 105-ft-long roof (32 m) (Barkow Leibinger Architects)

Fig. 5.17 Diagram of the tensions in the underside and topside of the roof (Barkow Leibinger Architects)

Fig. 5.18 Isometric representation of a segment of box girder (Barkow Leibinger Architects)

Fig. 5.19 Prototype of a segment of box girder (Barkow Leibinger Architects)

Fig. 5.20 Study of patterns representing the tensional forces of the roof (Barkow Leibinger Architects)

Fig. 5.21 Prototype of the cellular reinforcement of the roof (Barkow Leibinger Architects)

Fig. 5.22 The steel sheet was cut using a machining centre equipped with a laser (TruLaser 3030). The laser has 5,000 W power and allows cutting without introducing deformations and ensures smooth edges (TRUMPF GmbH)

Fig. 5.23 The steel sheets are folded into the final shapes by means of numerically controlled bending-presses (TruBend 3000). The exact bending angles were made automatically by means of free bending, which gives the ability to fold the sheet according to angles programmable at anything between 30° and 179° (TRUMPF GmbH)

Fig. 5.24 All the various processes of threading, forming, holing, and deburring are punched with a single stroke by the machine (TruPunch 3000). The threads are predisposed with the punching, which expedites assembly of the individual parts into the cellular structure (TRUMPF GmbH)

Fig. 5.25 Assembly in the shop floor of the segments to construct the lengthways girders of the roof (Barkow Leibinger Architects)

Fig. 5.26 (a, b) At the building site, the box girders were positioned and bolted together to assemble the complete honeycomb load-bearing structure, which was raised and hinged to the four columns (Corinne Rose)

Figs. 5.27 and 5.28 Randall Stout Architects. Plan of the fourth floor of the museum. Section of the atrium and the exhibition spaces (Randall Stout Architects)

Figs. 5.29, 5.30 and 5.31 The new regional Museum of Alberta rises up in the centre of the city of Edmonton, Canada (A. Zahner Company, Robert Lemermeyer)

Fig. 5.32 Cladding in the severe local stone accentuates the massiveness of the exhibition building, emphasised by the recessed joints (A. Zahner Company)

Fig. 5.33 Aurora borealis (United States Air Force Senior Airman Joshua Strang, Wikimedia Commons)

Fig. 5.34 The wide area of the atrium crossed and interwoven with the surface of the aurora borealis (Randall Stout Architects)

Fig. 5.35 Angel Hair™ panel detail of the engravings on the surface of the stainless steel made with numeric control water jet (A. Zahner Company)

Fig. 5.36 Assembly on site of the Angel Hair™ cladding (A. Zahner Company)

Fig. 5.37 The external surface of the aurora borealis in stainless steel highlights the nocturnal reflections, the internal surface in white-painted aluminium reflects the direct illumination of the hall, creating a soft low light (Randall Stout Architects)

Fig. 5.38 The contact sections between the hollow tubes are calculated exactly so as to precisely control the cutting. The aim is to obtain the best area of contact between the tubes, for sake of the welding (Empire Iron Works Ltd)

Fig. 5.39 Empire Iron Works proceeds to the arc welding of girders and hollow tubes in order to put together the different subframing structures into which the complete aurora borealis is divided (Empire Iron Works Ltd)

Fig. 5.40 Model in Tekla of the sub-framing structure (Empire Iron Works Ltd)

Figs. 5.41 and 5.42 On the building site the sub-framing structures are joined together and to the building by means of flanges and clamping bolts (A. Zahner Company)

Figs. 5.43 and 5.44 Model in Tekla of the structure and substructure (Empire Iron Works Ltd)

Fig. 5.45 The model of the skin in Rhino (Randall Stout Architects)

Figs. 5.46 and 5.47 The model of the structure and the cladding of the aurora borealis in Rhino (Randall Stout Architects)

Fig. 5.48 The model of the structure and of the substructure in Tekla (Empire Iron Works Ltd)

Fig. 5.49 The structure during construction (Empire Iron Works Ltd)

Fig. 5.50 The total station leads the precise fitting together of the elements of the structure during assembly at the shop floor. After the welding, the coordinates of the individual elements of the structure are surveyed in order to build the CAD model of the assembled structure (Empire Iron Works Ltd)

Figs. 5.51 and 5.52 The subsystems of links and panels of the aurora borealis, the joints between the principal and secondary structures (A. Zahner Company)

Figs. 5.53 and 5.54 Edge of the aurora borealis, details of the secondary structure and of the joints of the panels in the ZEPPS™ system (A. Zahner Company)

Chapter 6

Fig. 6.1 Independent clamp log band saw and log carriage, with electronically controlled hydraulic blade push and blade tensioning unit, hydrostatic feed with speed of 98 ft 5 in 7/64 per minute (30 m/min) (Angelo Cremona SpA)

Fig. 6.2 (a, b) Log debarker-ring type, for logs of between 6 in 19/64 and 31 in 1/2 (16 and 80 cm) in diameter The cutting tools and the scraping tools carry out a separate pressure to adapt to the log (Angelo Cremona SpA)

Fig. 6.3 (a, b) Numerically controlled work centre for the production of windows and doors (Biesse Group)

Fig. 6.4 Rotary slicer for making sheets of up to 158 in (400 cm) long. Includes a thermoregulation group for the tools in order to avoid stains on the veneer surface and for remote control of the cutting parameters (Angelo Cremona SpA)

Fig. 6.5 Bentwood Thonet chair dismantled for shipping. This Thonet model is considered one of the most successful pieces of furniture (Michael Thonet)

Fig. 6.6 Stretching press for bending items in solid wood of length varying between 2 and 118 in (50–3,000 mm) with maximum thickness of 3 in 1/2 (90 mm) (Vecchiato Valter & C. Snc)

Fig. 6.7 Peeling lathe for logs of large diameter with electronic control system for the machine and for the sheet (Angelo Cremona SpA)

Fig. 6.8 Drying plant for green sheet with automatic loading and unloading (Angelo Cremona SpA)

Fig. 6.9 Gluing line for sheets with automatic loading and unloading (Angelo Cremona SpA)

Fig. 6.10 Pressing machine for consolidating panels with automatic loading and unloading (Angelo Cremona SpA)

Fig. 6.11 Pressing machine for consolidating laminated beams with a maximum length of 105 in (32 m), height between 63 and 86 in 1/2 (160 and 220 cm). The pressure exerted along the beam is controlled and managed electronically, also to adapt it to beams of varying thickness (Vecchiato Valter & C. Snc)

Fig. 6.12 Semi-automated plant for the production of laminated panels of solid timber in cross-grained layers (X-Lam) (Acimall, the Italian Woodworking Machinery and Tool Manufacturers Association)

Fig. 6.13 The street-level with the shops and the market (Jürgen Mayer H. Architects)

Fig. 6.14 The public plaza 16 ft 4 55/64 in (5 m) above ground (Jürgen Mayer H. Architects)

Fig. 6.15 The balcony plaza at 70 ft 6 15/32 in (21.5 m) above ground and the 'architectural walkway' (Jürgen Mayer H. Architects)

Fig. 6.16 Longitudinal cross-section (Jürgen Mayer H. Architects)

Fig. 6.17 The city level with the shops and the market and the raised public plaza at 16 ft 4 55/64 in (5 m) above ground (Angel Vilches)

Fig. 6.18 The nave of Seville Cathedral, the third largest of the Christian world (Pom[2])

Fig. 6.19 Villard de Honnecourt elevation from the western tower of Laon Cathedral, from the *Livre de portraiture*, 1220–1235 (Paris, Bibliothèque Nationale de France, ms. fr. 19093, c. 10r)

Fig. 6.20 The grid of the parasol was generated on the digital model by removing from the organic forest a lattice of squares of 4 ft 11 in (1.5 m) side (Ignacio Ysasi)

Fig. 6.21 The steel composite structure of the restaurant platform at 70 ft 6 15/32 in (21.5 m) above ground level (Jürgen Mayer H. Architects)

Fig. 6.22 Kuka robot with seven axes, used by Finnforest Merk for the cutting, milling and drilling of each panel (Finnforest Merk GmbH)

Figs. 6.23 and 6.24 The nodes at the joining points of the structure must function in a wide range of angles, guarantee stability between the panels and be compact as they are on view (Finnforest Merk GmbH)

Fig. 6.25 Detail of the structural node made of the fork-headed connection. The coupling hooks are fixed to each other with steel pins. The flange of the fork is fixed to the panels using preloaded bolts screwed into the threaded rods (Finnforest Merk GmbH)

Fig. 6.26 The threaded rods are fixed to the panel by gluing in the holes made in the panel end (Finnforest Merk GmbH)

Fig. 6.27 The special fork-headed connection is bolted onto the rods of the panels (Finnforest Merk GmbH)

Fig. 6.28 The panel is lifted into the assembly position, where the fork-headed hooks are fixed in their final position with the insertion of steel pins (Finnforest Merk GmbH)

Fig. 6.29 Plan of the timber structure of the inner dome (Stratex SpA)

Fig. 6.30 The rods which join the metal structure of the outer dome and the timber structure of the inner dome with the interposition of acoustic insulators (Alberto La Tegola, Stratex SpA)

Fig. 6.31 The erection of the inner dome in laminated timber (Alberto La Tegola, Stratex SpA)

Fig. 6.32 Three-dimensional model of the laminated timber inner dome, the steel outer one and the brick drum (Stratex SpA)

Fig. 6.33 Three-dimensional model showing the tie roads between internal and external domes (Stratex SpA)

Fig. 6.34 Detail of the carpentry work of the joint between round windows and beams (Stratex SpA)

Fig. 6.35 Milling of beams in laminated timber through work centre in the Stratex plant (Stratex SpA)

Fig. 6.36 One of the 72 curved main beams, 16 m long at the level of the building site (Alberto La Tegola, Stratex SpA)

Fig. 6.37 Detail of the laminated timber structure (Alberto La Tegola, Stratex SpA)

Fig. 6.38 The structure of the inner dome completed before being covered (Alberto La Tegola, Stratex SpA)

Index

L. Caneparo, *Digital Fabrication in Architecture, Engineering and Construction,*
DOI 10.1007/978-94-007-7137-6, © Springer Science+Business Media Dordrecht 2014

Printed by Books on Demand, Germany